Digital Information and Knowledge Management: New Opportunities for Research Libraries

Digital Information and Knowledge Management: New Opportunities for Research Libraries has been co-published simultaneously as *Journal of Library Administration*, Volume 46, Number 1 2007.

Digital Information and Knowledge Management: New Opportunities for Research Libraries

Sul H. Lee
Editor

Digital Information and Knowledge Management: New Opportunities for Research Libraries has been co-published simultaneously as *Journal of Library Administration*, Volume 46, Number 1 2007.

Routledge
Taylor & Francis Group

NEW YORK AND LONDON

First Published by The Haworth Press, Inc.,
10 Alice Street, Binghamton, NY 13904-1580 USA.

This edition published 2012 by Routledge
711 Third Avenue, New York, NY 10017
2 Park Square, Milton Park, Abingdon, Oxon, OX14 4RN

Routledge is an imprint of the Taylor & Francis Group, an informa business

Digital Information and Knowledge Management: New Opportunities for Research Libraries has been co-published simultaneously as *Journal of Library Administration*®, Volume 46, Number 1 2007.

The development, preparation, and publication of this work has been undertaken with great care. However, the publisher, employees, editors, and agents of The Haworth Press and all imprints of The Haworth Press, Inc., including The Haworth Medical Press® and Pharmaceutical Products Press®, are not responsible for any errors contained herein or for consequences that may ensue from use of materials or information contained in this work. With regard to case studies, identities and circumstances of individuals discussed herein have been changed to protect confidentiality. Any resemblance to actual persons, living or dead, is entirely coincidental.

The Haworth Press is committed to the dissemination of ideas and information according to the highest standards of intellectual freedom and the free exchange of ideas. Statements made and opinions expressed in this publication do not necessarily reflect the views of the Publisher, Directors, management, or staff of The Haworth Press, Inc., or an endorsement by them.

Cover design by Kerry E. Mack.

Library of Congress Cataloging-in-Publication Data

Digital information and knowledge management : new opportunities for research libraries / Sul H. Lee, editor.
   P. cm.
   "Co-published simultaneously as Journal of library administration, volume 46, number 1."
   Includes bibliographical references and index.
   ISBN-13: 978-0-7890-3565-3 (alk. paper)
   ISBN-10: 0-7890-3565-0 (alk. paper)
   ISBN-13: 978-0-7890-3566-0 (pbk. : alk. paper)
   ISBN-10: 0-7890-3566-9 (pbk. : alk. paper)
   1. Research libraries–Collection development. 2. Academic libraries–Collection development. 3. Digital libraries–Collection development. 4. Libraries–Special collections–Electronic information resources. 5. Electronic information resources–Management. 6. Knowledge management. 7. Libraries and scholars. I. Lee, Sul H. II. Journal of library administration.
Z675.R45D54 2007
025.2'1877–dc22
                                        2006028645

Digital Information and Knowledge Management: New Opportunities for Research Libraries

CONTENTS

ABOUT THE EDITOR

Sul H. Lee is Peggy V. Helmerich Chair and professor of Library and Information Studies at the University of Oklahoma and dean of University of Oklahoma Libraries. Professor Lee has taught in the School of Library and Information Sciences and directs a major university research library with a collection exceeding 4.5 million volumes. He has served as professor and dean at the University of Oklahoma since 1978 and is the senior dean on the University of Oklahoma campus in Norman, Oklahoma.

Professor Lee's academic background is in political science, international relations, and library and information science and he holds graduate degrees in these disciplines. He is author of more than two dozen books in the field of librarianship, along with numerous articles and professional presentations. In addition to his current positions at the University of Oklahoma, Professor Lee has taught at Oxford University in Oxford, England and the University of Michigan. He is internationally recognized as a consultant on libraries and has served on important national and regional professional organizations and consortiums such as the Association of Research Libraries board of directors; the board of governors for the Research Libraries Group (RLG); the Council of the American Library Association; and as chair of the Greater Midwest Research Library Consortium.

He is also editor-in-chief for The Haworth Press's academic journal division and editor of Haworth's *Journal of Library Administration.* He serves regularly as consultant to library service providers, academic book vendors, publishers, and advises state and local governments on library affairs. His outstanding career spans more than 40 years in academic libraries and he has witnessed the transition of libraries from the era of card catalogs to the proliferation and general acceptance of digital information.

Introduction

Sul H. Lee

This volume addresses digital information and knowledge management and the opportunities that both offer to librarians. Knowledge management has been advocated and used in the business world for well over a decade now, and there have been several attempts to apply it in academic libraries. While there has been no clear consensus as to how knowledge management might work best in, and to the benefit of an academic library, electronic resources have given the 21st century library new roles to fulfull and created a demand for librarians skilled in the acquisition, retrieval, and dissemination of digital information. The advent of the digital library has presented librarians with an exciting new scenario for the profession. Instead of becoming superfluous, librarians and libraries have met the new challenges presented by digital resources and have moved from building collections of print materials into a newer field frequently referred to as knowledge management. The papers that follow explore the unfamiliar territory of knowledge management and offer insights into how academic libraries are moving from managing collections of print and digital resources into the broader spectrum of managing knowledge. Paula Kaufman, University Librarian, University of Illinois at Urbana, set the tone for the conference in her well-structured paper that outlined the challenges presented to traditional libraries by digital resources. She concluded that the digital information and knowledge management offer tremendous opportunities for the 21st century library.

[Haworth co-indexing entry note]: "Introduction." Lee, Sul H. Co-published simultaneously in *Journal of Library Administration* (The Haworth Information Press, an imprint of The Haworth Press, Inc.) Vol. 46. No. 1, 2007, pp. 1-3; and: *Digital Information and Knowledge Management: New Opportunities for Research Libraries* (ed: Sul H. Lee) The Haworth Information Press, an imprint of The Haworth Press, Inc., 2007, pp. 1-3. Single or multiple copies of this article are available for a fee from The Haworth Document Delivery Service [1-800-HAWORTH, 9:00 a.m. - 5:00 p.m. (EST). E-mail address: docdelivery@haworthpress.com].

Available online at http://jla.haworthpress.com
© 2007 by The Haworth Press, Inc. All rights reserved.
doi:10.1300/J111v46n01_01

While enormous amounts of new data and information are now available to us, Dennis Dillon, Assistant Director, University of Texas at Austin Libraries, sees increased confusion and complexity in the presentation of that knowledge. More does not always equal better and it falls to librarians to sort out, make sense of, and present coherent information to those who use libraries. It is a daunting task, but one that librarians must accomplish.

Sarah Michalak, Librarian and Associate Provost for University Libraries at the University of North Carolina at Chapel Hill, and her colleague Judith M. Panitch, Director of Library Communications at University of North Carolina at Chapel Hill, note that digital resources have bought a different approach to the way librarians present information. Librarians used to rely more on one-to-one consultations, but the digitization of collections has greatly widened the dissemination of information. They also observe that wider usage of digital collections generates a need to present a standardized method of information presentation that meets scholarly criteria. Their article compares the approaches of this task in two academic libraries.

Vice Chancellor for Information Technology and Dean of University Libraries, Washington University at St. Louis, Shirley K. Baker suggests in her presentation that libraries not only have the opportunity, but also the obligation to manage digital information and knowledge for their constituencies. She insightfully points out that the challenges of the digital library are less technical than cultural and financial. Her paper outlines some successful approaches to these problems and some failures.

Charles T. Cullen, a historian, a former editor of scholarly papers, and president of the Newberry Library, offers a scholar's view on the digitization of collections and historical materials. He advocates caution and sound planning for the digitization of collections with careful attention paid to the needs of scholars who will be using the materials.

President of the Council on Library and Information Resources, Nancy Davenport acknowledges the many changes and challenges that digital information and knowledge management have brought to academic libraries. She also observes that these dramatic and radical changes have also affected the way scholars in many disciplines view information resources. It falls to librarians, she concludes, to maintain the connections between academic disciplines and libraries.

In the closing article, Gary M. Shirk, President of Yankee Book Peddler, gives us an interesting and different perspective on how digital materials are changing our perception of information. Digital resources, he notes, offer us new ways to view large quantities of information while developing a topology of library collections.

This collection of papers from eight of the nation's leaders of academic libraries presents new viewpoints on digital information and knowledge management. As in years past, I am grateful to have the opportunity to publish them in both journal and monograph formats with the hope and expectation that they will be of use to the academic library community.

It's Not Your Parents' Library Anymore: Challenges and Opportunities in the New Webs of Complexity

Paula Kaufman

SUMMARY. The 21st century holds the promise of a modern Renaissance in which traditional research library collections and services collide with the promises and realities of digital information and knowledge management. The most intriguing challenges and opportunities for research libraries center on the juxtaposition between the traditional library and digital information and knowledge management, with resultant webs of complexity. This paper examines these challenges and opportunities and raises questions of values and principles, missions and strategies, content and copyright, and methods and ethics. doi:10.1300/J111v46n01_02 *[Article copies available for a fee from The Haworth Document Delivery Service: 1-800-HAWORTH. E-mail address: <docdelivery@haworthpress.com> Website: <http://www.HaworthPress.com> © 2007 by The Haworth Press, Inc. All rights reserved.]*

KEYWORDS. Knowledge management, challenges to libraries, societal change, universities and digital information, print collections, digital collections, library services and digital information, change and academic libraries

Paula Kaufman is University Librarian, University of Illinois at Urbana, 230 Main Library, 1408 West Gregory Drive, Urbana, IL 61801 (E-mail: ptk@vive.uiuc.edu).

[Haworth co-indexing entry note]: "It's Not Your Parents' Library Anymore: Challenges and Opportunities in the New Webs of Complexity." Kaufman, Paula. Co-published simultaneously in *Journal of Library Administration* (The Haworth Information Press, an imprint of The Haworth Press, Inc.) Vol. 46, No. 1, 2007, pp. 5-26; and: *Digital Information and Knowledge Management: New Opportunities for Research Libraries* (ed: Sul H. Lee) The Haworth Information Press, an imprint of The Haworth Press, Inc., 2007, pp. 5-26. Single or multiple copies of this article are available for a fee from The Haworth Document Delivery Service [1-800-HAWORTH, 9:00 a.m. - 5:00 p.m. (EST). E-mail address: docdelivery@haworthpress.com].

Available online at http://jla.haworthpress.com
© 2007 by The Haworth Press, Inc. All rights reserved.
doi:10.1300/J111v46n01_02

We're in the business of giving away knowledge. For free. Come in, please come in, and take some knowledge for free, no, no limit, keep going, gorge on it if you want, no, it's not a trick, a come on, a free sample and then we'll bill you later, or we'll paper your head with banners and pop-ups. Librarians don't have a lot of status and we don't make a lot of money, more than poets, but not so much, say, as your more successful panhandlers, so our ideals are important to us and the love of books and the love of knowledge and the love of truth and free information and letting people discover things for themselves . . .

–Larry Beinhart, *The Librarian*[1]

INTRODUCTION

Giving the last guest lecture to a first-semester class of library school students is always pleasurable and sometimes adventurous. The students have plowed through challenging readings, engaged in stimulating discussions, and prepared several portfolios and they have a smattering of knowledge and understanding of the issues faced by contemporary libraries that make them curiously self-confident and thus able to ask some extremely interesting questions. Recently, a very serious young woman asked how I could expect contemporary academic public service librarians to understand all the complexities of online catalogs when all I had needed to know thirty-five years ago, when I began my career as a reference librarian in an academic business library, was the simple card catalog. Her question was not entirely naïve, but it was provocative. The card catalog was not a simple tool and one had to understand its complexities, that is, the intricacies of catalog records, name and subject authorities, and idiosyncratic filing rules (not to mention what to do if a card was misfiled or you dropped a tray of catalog cards with the rod removed). I did not mention that the enormous filing backlogs, which were one of our "dirty little secrets," made finding many items, particularly newly-accessioned materials, impossible. The card catalog was the primary finding tool available in libraries for most of the twentieth century; it was supplemented by printed catalogs, bibliographies, indices, abstracts, and other finding tools. They were fixed and separate and could only be used one at a time and by only one person at a time.

The information world into which this young woman will step soon as a librarian is quite different, and it is clear that tomorrow's world will be even more complex. She will perform many of the same functions that I did. She will help users find what they want and she will teach them–one-on-one and in groups, face-to-face and virtually–to find and evaluate information themselves. Further, she will create tools to facilitate their searches. But she will do this in a world of incredible numbers of information access tools and information resources–webs of complexity created not by the transformation of tools from paper to digital form but by their very integration into the universe of information–and she will be confronted by the demands and expectations of an on-demand generation, grown up in the digital age, who demand instant access, delivery, and control. The transformation of the carbon-based world to one that is increasingly silicon-based will provide her the challenges and the opportunities and the choices that I did not confront thirty-five years ago. This transformation and how it forms different cultures on different campuses will cause each library to make different choices that will result in them looking less and less like one another.

"The technological excursions of recent decades have advanced societies in which silicon and carbon, and the systems they generate, permeate our lives and weave webs of complexity that will profoundly challenge the way we live and how we see ourselves and relate to each other, locally and globally."[2]

In 2002/2003, the University of Illinois at Urbana-Champaign undertook a year-long exploration of the anticipated intersections and interactions of silicon, carbon, and culture through the arts, humanities, and technology. The initiative, which was undertaken jointly by the College of Liberal Arts and Sciences and the College of Fine and Applied Arts, with support from the Madden Initiative in Technology, Arts, and Culture, and the University's Chancellor and Provost, explored the interplay between the arts, humanities, sciences, and technology fields at UIUC. Projects sponsored by the initiative explored such topics as:

- Visualizing the Global: New Knowledges, Cyber-Globalization, and a Reorientation of Perspectives
- Memory and the Construction of Identity and Culture
- Hands On, Plugged In: Living on the Prairie
- "Walking" Through Knowledge Networks in Virtual Space

This initiative represents what is perhaps the most important challenge of the twenty-first century: the implications of the juxtaposition between silicon and carbon. Silicon and carbon can interact in pervasive and sometimes terrifyingly invasive ways, making the body increasingly technological and technology increasingly human-like. Contemporary ideas of private spaces and public interactions inevitably will be challenged. It is these interactions that epitomize the essence of the transformation we are experiencing in academic research libraries and it is these interactions and transformations that our young student will have to face and shape as her career matures.

The global e-future looks as dangerous as it does adventurous. Today, the outlines of this future are being drawn by scientists and engineers. But unless philosophers, writers, artists, social scientists, performers, public intellectuals, policymakers, librarians, and others join the design process, the silicon future will be less rich than the carbon one from which we are emerging. As the barriers of time and space dissolve, artists and scientists and librarians and others must interact with each other in their own spaces–in laboratories and studios, offices and libraries–however these spaces change, to rediscover the energies of learning together and innovating collaboratively.

Libraries have long been represented by tangible symbols: the jewel in the university's crown, the heart of the university, the campus's treasure. These images are remarkably similar from campus to campus. Our large main library buildings are important iconic representations of our place in the university. But these images are static; they connote traditions–often described as supportive of teaching, learning, and research–that are grounded in carbon at a time when the world in which we operate is transitioning to one dominated by silicon. To represent our future, Wendy Lougee uses the clever metaphor of fabric–a commodity that spreads across the institution flexibly, fluidly, and dynamically, where and when it is needed.[3] The image of fabric represents the library as a more integral, integrated, and collaborative commodity than does a jewel or a building that stands fixed and alone. Each piece of fabric is distinguished by different textures, colors, shapes, and sizes, just as each academic research library will become increasingly distinguished by its diffusion, its shapes, its expertise, and its services as the twenty-first century progresses.

Every generation of librarians has met grand challenges. After struggling through the financially challenging Depression years, my parents' generation seized the opportunity to build great collections by scooping up materials from bombed-out libraries and private individuals whose collections had been ravaged by war and whose needs for cash were greater than their need to keep their books. These materials, coupled with the huge flow of materials made available through such post-WWII programs as those supported by PL480, produced masses of books that could not be processed for many years, resulting in enormous backlogs of inaccessible materials. That generation also built great buildings, established systems of departmental libraries that were located where disciplinary faculty were located, and reacted to its inability to provide on-site everything their clienteles needed by shaping interlibrary loan systems and laying the groundwork for some of the major collaborative institutions on which my generation has relied. In fighting for academic freedom in the McCarthy era, my parents' generation codified values that were most important to maintain.

My generation rose to the challenge of developing new bibliographic standards, new best practices, new technologies, new means of providing access, and rising expectations and demands for access to content and services that crossed the boundaries of time and place. At the start of our careers, we almost always had sufficient resources, but that did not last long. One of the first papers I ever presented was entitled "How To Say 'No' and 'Why' Diplomatically."[4] A significant proportion of the funds that universities had traditionally made available to libraries were diverted to support burgeoning campus information technology operations, creating enormously challenging declines in support for library collections and staff. At the same time, increases in scholarly output, the penetration of commercial publishers into the marketplace, new publishing methodologies, rising prices of scholarly materials, the growth of big science on our campuses, and dramatic rises in general inflation rates coupled with a consequent fall in the value of the U.S. dollar, created enormous losses in buying power and caused us to begin the first wave of cancellations of large numbers of serial titles and to decrease spending for monographs in an effort to try to support "core" scientific journals. Our collections became more and more like each other's.

At the same time, the causes and extent of the previously unexplained deteriorating physical state of many of the materials in our collections became known and we established programs to improve environmental conditions and preserve the intellectual content of materials too fragile to save physically. We became advocates and activists in trying to release the bonds the commercial sector held over much of the production of scholarly communication. The reintroduction of important information public policies that created obstacles to access and privacy–copyright and filtering, for example–presented additional barriers to our success; we rose to the challenge by becoming involved actively and sometimes successfully in trying to change public policies. We continued to reaffirm the profession's core values as we continued to meet these challenges.

Many of the choices we made to improve content availability and services to users were successful because we worked together or because they were obvious or easy to make. We adopted the MARC standard and began to use OCLC's services; introduced online catalogs, first to improve the efficiency of cataloging and then to improve services; developed metadata schema and harvesting tools to improve access; cancelled serials; turned to microfilm both to save shelf space and to preserve the contents of the materials we finally recognized were deteriorating on our shelves and we persuaded the federal government to support some of our efforts. As a result, the libraries we ran all bore a great resemblance to one another. They still do. Organizational cultures may differ from one institution to another, but when any of us walks onto another campus, we know fairly well where the library is, how it is configured, and what range of services it provides. We still know many libraries by the strengths of their locally-owned collections–at least for now.

My generation, however, has not until now faced the grand challenges presented by the transition from carbon to silicon and the fundamentally different conceptualizations of the library's and the librarian's role within the academy it presents. The choices each library makes today and tomorrow will differ greatly one from the other. Many choices will not be easy or obvious and many will not rely on collaborative action. Being strategic will be very important; being trendy will no longer be important at all. The libraries that your children's generation will take over from you for the most part will look very different from today's libraries, and most of them will not resemble each other nearly as much as today's libraries do. There are adventures and dangers ahead. How one assesses the risks and what choices one makes will determine which of this generation's libraries will thrive, which will decline, and which will disappear.

Rather than try to articulate a fully developed vision of the future, I want to focus on a few observable trends and the challenges and opportunities they present. What is changing around us in society, in universities, and in libraries? Where will the challenges and opportunities lie? What choices are there to be made? Those are some of the questions on which this article will center.

SOCIETAL TRENDS

We live in an 'on-demand' world that centers increasingly on 'me.' The Internet has become the most important source of current information for most people today–the primary place they go for research, general information, hobbies, entertainment listings, travel, health, and investments. We can expect that the Internet will evolve to become much more universal, ubiquitous, and pervasive than it is today. Although we can predict that all media will eventually move around the world in tiny packets that will be basic units of tomorrow's communications, we are not yet certain of its importance.[5] Visions of the future painted by futurists such as Ray Kurzweil,[6] who envisions a world where the difference between man and machine blurs and where the soul and the silicon chip unite, may not emerge by 2020 as Kurzweil predicts, but it is clear that the Internet will become increasingly ubiquitous and important.

We are learning some very interesting things about Internet users. The more experience users have with the Internet, the less television they watch. The ramifications to a nation whose population once spent a large portion of time in a passive activity (television) and now transfers that to an interactive activity (the Internet) are profound. It could affect every aspect of American culture, the economy, politics, and social behavior. And it will likely affect reading habits, with people moving from print to online publications and from static to dynamic documents. The increasing access to full text documents, exemplified by Google's new undertakings that will provide access to scholarly materials[7] and that will digitize millions of printed books in the next decade[8] will have profound effects on how people access and use information and will further blur the boundaries between 'scholarly' and 'popular' works. The ubiquity of information on the Internet, coupled with the ability to access it in an instant, raises expectations and demands that librarians will need to meet independent of time and place. That is certainly something librarians of my generation did not have to consider a generation ago.

Many aspects of society are being changed by the transition from a carbon to a silicon-based world. Perhaps nothing exemplifies this better than mass marketing, which has been stood on its head by the shifting emphasis from selling to the vast, anonymous crowd to millions of specific consumers, has particular importance to academic librarians. For marketers, the evolution from mass to micromarketing represents a fundamental change driven as much by necessity as opportunity. Further, the proliferation of digital and wireless communication channels is diluting yesterday's mass audience that was accessible through a handful of media outlets to individual audiences spread across hundreds of cable-TV and radio channels, thousands of specialized magazines, and millions of computer terminals, video-game consoles, personal digital assistants, and cell-phone screens.[9] The relentless search for products and services that are "right for me" will continue to drive demand. Ultimately, this is all about offering a degree of customization for everyone, everywhere. The same technological advances that are fragmenting the mass audience are also empowering a new class of digitally savvy consumers who compile, edit, and otherwise customize the media they consume to their own personal requirements. Clearly what were the common experiences of my generation are evolving into the more individualized experiences of members of the current generation.

A recent article in *Business Week*[10] points out some of the implications of these new societal behaviors. My generation hailed the remote control, which began to give us some easier control over what we watched. Who has not channel-surfed or muted commercials? The remote control was just the tip of the iceberg, however. While members of the millennial generation consume gobs of digital fare, they also have mastered technical tools to evade marketers and to customize their own programming. Customers, with their fingers on the delete button, the mouse, or the remote, wield control as never before. Instead of blasting consumers with fusillades from the TV, advertisers must reach them through hundreds of web sites, TV and radio channels, video games, music downloads, and more.

Members of the millennial generation are also renowned for their multitasking. Ninety-nine percent of college students use e-mail, and 59 percent use instant messaging. The first generation to grow up clicking mice, college students are used to controlling the flow and content of programming, whether music downloads or TV. And there appears to be fervor for up-to-the-second information. Members of this generation will spend money, but they demand options and control. And control over their seemingly unending range of choices gives them unprecedented power.

How the world has changed as we have moved from mass to micro! Old consumers passively received network broadcasts; new consumers are empowered media users who control and shape content. Old aspirations were to keep up with the crowd. New aspirations are to stand out from the crowd. Old brands were big and ubiquitous; new brands are niche brands and product extensions, and mass customization means many new variations.[11] Old library users came to a building to use or borrow materials; new ones access content independent of time and place, fettered by the geography of a single physical collection.

Public policies have always been important to the flow of ideas, information, and knowledge. The global place of the United States, the specter of fear of enemies from abroad and from within, and philosophies that position our freedoms in current contexts have varied from era to era. Formers and current generations fended off threats to academic freedom in the McCarthy era, threats to privacy in the Cold War era, and threats to the free flow of information as new copyright policies emerged in the age of the photocopier. Recent global developments in copyright protections in the emergent digital age and reactions to the terrorist attacks in the United States on September 11, 2001 have created current policies that reflect underlying philosophies of protection for creators and greater governmental powers to intrude on what had been interpreted previously to be private activities. As the United States struggles to define its place in a world it no longer dominates and in which it is no longer the center of higher education to which the rest of the world is drawn, coming generations will undoubtedly struggle with policies that threaten the fundamental values of librarianship. Without aggressive advocacy, huge amounts of content and relatively unrestricted access and demand for control over content on the one hand, and increasing intrusions of governmental entities into access and use on the other, are set to clash–but hopefully not to crash and burn–in the coming years.

THE UNIVERSITY CONTEXT

Research universities are unique institutions, defined by their underlying mission to generate and disseminate knowledge in all spheres. They are perhaps the only institutions that bring together scientists, scholars, and artists to carry out this work themselves and to transmit the values and tools of their fields to the next generation. The challenges and choices academic research libraries will confront cannot be understood fully without understanding current and anticipated changes in research universities and the cultural changes that shape institutional growth.

U.S. colleges and universities are facing many challenges as they make their way through the early years of the twenty-first century. Financial constraints (which in public institutions are causing shifts from dependence on state funding to dependence on tuition and private fund-raising), coupled with profound changes in electronic and networking technologies, new interdisciplinary collaborations, increasing dependence on 'big science' grant funding and technology transfer, the withering of the vast pool of international students, increased focus on teaching undergraduates in research universities, and changes in scholarly communications, are just some of the forces that are creating conditions for changes as deep and important as those experienced during other transformative periods. These are exemplified by the Morrill Act of 1862, which led to the establishment of land grant institutions; the rise of graduate education and the development of disciplinary-based departments in the late nineteenth and early twentieth centuries; the GI Bill, which created new and unprecedented opportunities for the democratization of higher education and new expectations from the system; and the rise of federally-funded "big science," that grew during the post-WWII, cold war era.

Public universities continue to have a mandate to educate the majority of America's higher education student. The states that traditionally funded them, now faced with increasing costs of health care, security and law enforcement, and K-12 education, are increasingly unable to provide the levels of funding their compact with higher education requires. Previous plunges in state funding in the 1980s and 1990s were fairly quickly erased by rebounds in the nation's economy that were mirrored in most states, and although unfunded federal mandates, most notably in health care, put large liens on state budgets, rapidly rebounding revenues outpaced these commitments. But the tale of the first decade of the 2000s has so far had a different story line. Large state commitments and more unfunded mandates, coupled with a reluctance to impose new taxes, even–or perhaps especially–at a time when federal tax rates were being reduced, have led to cuts in funding for higher education and widespread acceptance that even when state economies rebound, increased funding for higher education is very unlikely. These financial challenges are troubling and appear to be permanent, and they are not unique to public institutions; with the exception of the most elite private colleges and universities, many private institutions are also struggling financially. Just as we are facing these dire budgetary circumstances, we are developing some of the most promising models for teaching, learning, student engagement, delivery of information content, and the use of technology.[12] Our challenge will be to take advantage of these models while operating in an environment of declining financial resources. Fundamental changes will characterize those institutions that will thrive throughout the twenty-first century.

At the same time, the world of scholarly communication, the arena in which information and knowledge are created and disseminated, is undergoing an extraordinary transformation, a transformation of similar–or perhaps even greater–importance as the invention of the printing press or the development of scholarly societies. Technologies such as e-mail have changed fundamentally the ability of scholars and students to collaborate across time and place and the computing environment is being restructured fundamentally on most campuses by the deployment of new classes of systems such as portals, learning management systems, and institutional and disciplinary repositories. "E-mail, electronic discussion lists, Web sites, and other communications platforms have offered opportunities for new and joint efforts of many different kinds to achieve results not even imaginable just a few years ago. Yet, even as these changes have had a substantial impact on many of the primary activities of scholars and students, the challenges of implementing the more profound changes in the system are far more significant and difficult to overcome."[13]

Wendy Lougee argues that current trends of distributed computing and open networks, coupled with emerging models for scholarly communication, have eased the boundaries among stakeholders, which in turn allows more permeable and overlapping of roles. "Content once fettered by physical constraints has been loosened. The conventions of scholarly communication have been stretched and opened to a wider audience. The products of publication have become more process-like. The roles of libraries have also changed to embrace new opportunities for facilitating and shaping content, communication, and collaboration."[14]

Although transformations from carbon to silicon have enabled scholars to collaborate more easily with each other, the products of scholarly communication have not yet changed substantially in form. Authors still make arguments and convey information, ideas, and insights. However, those changes are on the horizon. Some new genres of scholarly communication are beginning to emerge. The University of Virginia's Institute for Advanced Technology in the Humanities, for example, is beginning to produce a new genre that expands the capabilities of the traditional monographic form. In the coming years, we can expect to see scholars who have grown up from the very beginning of their careers working in digital environments, creating new forms and formats, multimedially integrating works that diminish the privileged place that text has historically occupied.[15]

Silicon has permeated many areas of our traditional universities. Many have invested hundreds of millions of dollars in enterprise-wide administrative and student information systems, designed to improve the efficiency and effectiveness of management processes. However, few of these implementations have been trouble-free; although some of the barriers to success are attributable to the technology, a much larger part may be attributable to the insufficient attention paid to organizational culture–processes, politics, and patterns of information sharing, or lack of sharing.[16] These experiences are extremely informative to those leaders who will be transforming today's libraries into tomorrow's, for understanding, managing, and successfully changing organizational culture is at least as important as technology to the ultimate goals of any enterprise.

In addition to trying to change organizational culture (overtly or not), university leaders are also trying to change the culture of how scholarship is done on their campuses. "Silicon, Carbon, and Culture" is but one of many examples of strong encouragement for faculty in different fields to collaborate in new ways and in new places. However, many scholars do not need such encouragement from university leaders. Changes in disciplinary and interdisciplinary scholarship are deep, complex, long-term, social, and cultural, and they are fundamentally reconfiguring the university's core activities. Although a stable infrastructure is essential to enable scholars and disciplines to create scholarly content and the tools that serve the need of the discipline, the use of the infrastructure and the tools required, and thus the training required for scholars in the field to use them, will differ from one discipline to another. More intractable than the technological issues are those of changing organizational structures, defining new roles for scholars and publishers and incorporating innovation into traditional environments,[17] and changing organizational culture.

IMPLICATIONS FOR ACADEMIC RESEARCH LIBRARIES

What does it mean for those who lead, manage, and work in academic research libraries when tradition collides with digital promises and digital realities? When new generations of students have grown up in a 'me'-centered world of digitally-enabled power and control? When boundaries of time and place no longer exist? When scholars work in new collaborative forms and develop new genres, formats, and models of scholarly communications? When universities are facing unprecedented financial challenges and when their focus on fund-raising threatens to overshadow their focus on mission? When changes in public policies create an environment so severely restrictive that information cannot be sought freely or used without the specter of Big Brother looking over a virtual shoulder? And when organizational cultures both of universities and scholarly communication stakeholders are unsuited for the future yet appear to be impervious to change?

We are now at a critically important crossroad characterized by profound transformations from carbon to silicon. The arena in which ideas are created, shared, and documented, which is the world in which academic librarians operate, is undergoing a transformation of unusual scale and impact, equivalent or perhaps more phenomenal than the invention of the printing press. Content once fixed in ink on paper, bound and shelved one by one, has become unfettered and is now available to a broader audience that wants it just as it needs it, and wants to control it at any cost; as a result, the focus in libraries is shifting from inanimate collections to specialized expertise.

Yesterday's libraries were defined primarily by their collections, whose range, depth, size, and character were generally well known to scholars. Today's libraries are best characterized as a combination of collections, content, and expertise, but they still resemble one another greatly and they still operate, for the most part, as separate organizations within their universities. As libraries' general collections become more and more alike, 'special' collections–however defined–will take on increasing importance. Most libraries will diminish their purchasing of print and will license commonly available digital resources. Some libraries will continue to retain their current print collections, while many will rid themselves of the albatross these large print collections represent through digitization or deaccession, relying on a handful of libraries that are committed to investing the necessary space and financial resources to maintaining them for future 'just-in-case' use.

As increasing numbers of libraries diminish their print holdings, either actually or in relation to their increased access to digital content, libraries and librarians will begin to operate in ways that epitomize the fabric metaphor Wendy Lougee has evoked; the most vibrant libraries may not be seen as separate institutions at all. Librarians and the content to which they provide access literally will be diffused and infused into the University–in departments, laboratories, learning communities, learning systems, nooks and crannies, and more–in ways that we can barely imagine today. In fact, it is difficult to evoke a common visual image, a single symbolic representation, for unlike the libraries of the twentieth century, each of which bore remarkable resemblance to one another, it is likely that by the end of the century, no two academic libraries' array and delivery of services will be alike. As the century unfolds, and as libraries' collections become increasingly homogeneous, libraries' service array and delivery will diverge, their 'special' collections will become increasingly important, and they will bear less and less in common. Librarians will be path builders, but the paths they break will be quite characterized more by their divergence than their similarity.

Just as it is hard to think about a jewel morphing into fabric, or carbon morphing into silicon, it is no longer valid to think about our past and present morphing linearly into our future. Although we may think there have been parallel changes in the past, I believe that there has been sufficient discontinuity that we can no longer think incrementally. A decade from now, just like our fabric, many, but not all, libraries are going to be much more integrated into the university than they are today. Librarians will be skilled professionals who understand and work comfortably, integrally, and collaboratively within a wide range of disciplines and among the mass of newly developed and developing interdisciplinary areas that will come to dominate our universities. The library of the future will be about enabling the quest for knowledge through carbon and silicon in ways that will be characterized much more aptly as "collaborative" than as "supportive."

During most of the twentieth century, librarianship took shape by continuing to build collections of tangible materials and developing systems of access and services that mediated between individuals and content to serve expressed information needs. These systems generally treated all materials with the same descriptive schema, and the library's actions had little effect on the structure or functionality of published works. Thus, in general, the library was an organization that served all disciplines with similar tools and it served it in generally the same ways on most campuses, and it was largely an institution that operated separately from the creative or communication processes of other stakeholder groups.[18]

When it first became possible to publish electronically, the reference tools on which we rely became digital replicates of their printed versions. But, just as we have learned that just digitizing pages is insufficient to reconceptualize what they are in a digital networked information environment, so we need to think deeply about the new genres of scholarly output and the new genres of learning materials that will be produced by the new generations of course management systems. As do others, Clifford Lynch calls for libraries to invest and build partnerships with faculty and with creative explorers of this exciting yet dangerous environment, building institutional repository-like facilities, and trying to help people think through the structure of information in those settings.[19] Within this century we will witness an enormous sea change in the ways in which academic libraries are distinguished from one another. Scholars who grew up in the digital age are beginning to move from simple uses of a word processor toward a world where videos, images, sound, data and interactive materials are considered to be on an equal footing with text, or preferable to text and where traditional content will no longer hold its historically privileged place. As Lynch points out, this will fundamentally change what libraries do. "It is going to change. . . . relationships among scholars, people who know about information management, and libraries, publishers, and authors."[20]

How libraries make choices will be a fundamentally critical issue for each one of them, as well as for members of this and future generations of librarians. The commercial sector may provide some useful models. We have already seen many librarians shift roles from keepers of materials to managers of access to materials and providers of a wide range of value-added services, such as publishing or technology development, that had not been thought to be an important traditional part of a library's role. It is more these services and how they are delivered than it is their collections that will determine each library's future relevance and success.

Firms such as Amazon have taken advantage of their understanding of societal trends to offer an array of successful services, such as suggesting recently released or available materials based on filed profiles or past purchasing behavior. Firms such as Amazon have also capitalized on their technology applications and the convenient shopping they make possible to push the boundaries of a traditional "book store." Major search technology providers offer another useful model as they jockey hard for position on the desktops of individuals both in personal and work settings. Localization initiatives by major portal firms are repurposing all-purpose content for local markets.[21] These firms' future depend on responding to the needs of the 'me first' always-on, demand and control needs of today's generation and to whatever needs subsequent generations will express, and to continuously offering new services and acquiring or developing businesses once thought to be only peripheral to their core businesses. Many libraries will engage in these same behaviors. Some librarians will understand and take advantage of these trends to create an array of information services previously unavailable from academic libraries. Some libraries will expand into new niches and spaces. Some services may be so laden with additional value that libraries will offer them to users willing to pay for them. Unsuccessful libraries will remain mired in a carbon-intensive environment while the silicon-based world expands.

The rise of a silicon culture both in and outside of universities will continue to shape the choices each library will make; it will also help determine what functions will be important to each individual institutional setting and how they will be best delivered. Although "one size fits all" generic approaches will not be appropriate for successful libraries, we can make some generalizations nonetheless about the range and types of services that we might expect twenty-first century libraries to offer.

Today's libraries are much more serious about offering support for teaching and learning than those in which I started my career. We have progressed from giving tours and pointing out such features as "here are the index tables" to explaining the major disciplinary resources and their interrelationships to teaching students how to think critically about their information needs and how to find and evaluate the information they need. For the most part, however, library instruction sessions are offered on a course-by-course basis, often as an "extra" session. Similarly, although reserves services have moved from carbon to a mix of carbon and silicon, for the most part they deliver assigned readings to students in systems that are separate from the course management or other systems students and faculty use regularly. In the future, many libraries will choose to integrate information fluency instruction into course management systems, develop mass customized path-finding services pushed to students, and offer an array of classroom support services that integrate class readings, information instruction functions, and access to and delivery of content in all media and expertise into the systems the students and their teachers will be using characteristically. As more and more students become equipped routinely with devices that enable them to access library content and services ubiquitously, current "cutting-edge" information commons will be transformed from banks of computers and expert assistance available in library spaces into systems of access and expertise offered throughout campus.

Academic librarians have a long tradition of providing valuable research assistance, but until now it has been characteristically passive, and ill-named. Who, except for librarians and knowledgeable users, know what "reference" is? Yet today's reference services represent an enormous investment in personnel and content and they present users with a dazzling–and confusing–array of access choices. One can make an appointment, spontaneously walk up in person to either an "information desk" or a "reference desk," phone, send e-mail, or chat interactively with a reference staff member. With the exception of e-mail, all other access modes are limited to times in which the library is open and staff is available to help. In large academic libraries, characterized by many department libraries, the choices of where to go and whom to contact for help are impossibly confusing.

Choices about how to assist faculty and students differently than yesterday's models abound. Experiments with other models, such as the deployment of subject specialists to provide a broad range of assistance where faculty and students do their work–in laboratories, in departmental settings, on rounds with doctors, in residence halls and learning communities–or development of research commons designed to provide individualized and mass customized assistance for faculty and students who are doing research or 'pushing' content *a la* Amazon, are suggestive of new reference service paradigms in which librarians provide proactive research collaboration processes and infrastructure. Also intriguing are models in which librarians are playing stronger roles in evaluating information quality. Despite Stanley Wilder's recent and somewhat retrograde plea to abandon information literacy programs in favor of instruction in reading and writing scholarly material delivered at traditional reference desks,[22] these other newly-emerging models are much more compelling and perhaps signal the demise of the "traditional" reference services that were developed throughout the twentieth century.

Changes in the library's array of services, content resources, and products suggest concomitant changes in the spaces libraries occupy. They also suggest that investments in space will need to grow and that agility in the capacity to reconfigure or cede space for other uses will be paramount to success. Some libraries will continue to invest in large physical spaces while others invest in diffused spaces throughout campus. Most will choose a hybrid model–at least for the next few decades.

Although relatively simple to make, decisions about space configuration and use will be much harder and more complex to implement. Although librarians will have the opportunity to make choices about the configuration and ambience of their spaces, very persuasive arguments will be needed to secure the funds necessary to make changes. However, space will continue to be an important factor in a library's success, whether it is because of the ways in which the space is offered or because it is a financial drain to operate in an environment in which few or no acceptable alternatives are available. It is likely that future generations of faculty and students will confront a wide variety of library spaces that meet their needs increasingly less frequently.

Universities have always competed with one another–for faculty, for students, for grants and gifts, for prestige, for almost everything important to them. Libraries, while reflective of their parent institutions' competitiveness, have been more willing to collaborate and share resources to meet their primary goal of improving access to content and services to their campus communities. The 2003 OCLC environmental scan notes that "(s)sustainability is only possible through collaborations."[23] However, although collective action seems still to be preferable to individual action, as witnessed by the vitality of such initiatives as ARL's Global Resources Network, the Digital Library Federation's Aquifer project, and the emerging development of Ithaka's shared services, the Golden Age of Collaboration may be over. We may now be witnessing the beginning of a shift in the nature of collaboration. As predicted several years ago, the littered landscape of consortia can no longer be sustained;[24] we are seeing the start of a trend towards larger consortia as smaller ones come together to be more effective.[25]

And yet, competitiveness seems to drive much of the current divergence in the activities of libraries on a scale that is more than insignificant. Significant changes seem happening, however, as some libraries create pacts with the commercial sector instead of the library or public sector to make some of their contents available. Although most twentieth-century attempts to integrate library and computing operations in large institutions have been abandoned, many new and divergent activities are emerging. Stanford's High Wire Press initiative, the growing dependence on a small number of very large libraries to continue to collect and archive print resources and the different roles that traditional libraries are taking on in the development of digital libraries are only a few examples.

Academic librarians will continue to be confronted by profound financial challenges, emerging pockets of new organizational cultures within their institutions, and concomitant changes in demands and expectations for information resources and services. However, libraries and universities do not have a strong tradition of developing new services or products quickly. In a recent report, the National Research Council's warning to universities is equally applicable to libraries: "(P)rocrastination and inaction are dangerous courses during a time of rapid technological change. Universities will have to adapt themselves to a radically changing world while they protect their most important values . . . Although it is very difficult to predict the impacts on human behavior and other social institutions with any precision, higher education must develop mechanisms to at least *sense* potential changes and help it understand where the technology might drive it. . . . (S)trategies should include: the development of sufficient in-house expertise to track technological trends and assess various courses of action; the opportunity for experimentation; and the ability to form alliances. . . ."[26]

Models outside of academia might provide good paradigms. Many corporations look to fast-track a better consumer experience by using a five-step process: observation, brainstorming, rapid prototyping, defining, and implementing.[27] Others extend their models beyond content collection towards a more selective and distributed array of services designed to provide content value to their customers. The extension of weblog newsreading software to provide content aggregation services on the desktop in 2004 is a good example.[28] Librarians can argue about the lack of resources and about the restraining nature of many of their cultures, but if they wish to make changes as quickly as their users demand and if they want their libraries to be nimble organizations, they must view themselves as catalysts, as proactive change agents, and they must find ways to shed traditions and, while clinging to their values, make faster and better changes.

CONCLUSION

OCLC's recent environmental scan raised a most intriguing question: "What if libraries. . . . erased the organizational charts, the artificial separations of content, the visible taxonomies, and the other edifices real or otherwise built to bring order and rationality to what we perceive as a chaotic universe. What if we built an infosphere rich in content and context that was easy to use, ubiquitous and integrated, designed to become woven into the fabric of people's lives; people looking for answers, meaning and authoritative, trustable results. . . ."[29] What choices librarians make about whether and how to create integrated information communities that will serve scholars and enable the processes of tomorrow's scholarly communication and the decisions they make about how to best exercise control to add value and serve as catalysts for change will determine their futures. The choices are more abundant than ever before, and what librarians build will be determined by the choices they make.

The leaders of tomorrow's libraries have an array of important choices to make. Their choices will be dependent on a number of factors, including institutional aspirations, organizational cultures, changes in scholarly communications modes, new opportunities, changing values, public policies, and risk tolerance. In the future, a few libraries will be distinguished by their archives of print collections, some by their stewardship of the university's digital archives, some by their array of services, some by their integration into the learning environment, some by their integration into the research environment, some by their spaces, some by the absence of physical space. All will have some elements of all these characteristics and more, but no two late twenty-first century libraries will be alike.

The current and future generations of librarians have challenges and opportunities of unprecedented import. Clearly, contemporary libraries are not mere repositories. They are physical and social institutions, temples to our collective culture that reflect the high value we place on the information they preserve for us.[30] As Janet Murray wrote, the promise of silicon is that it "will make valuable information available to more people with less effort. If we can figure out how to do that we will not have obliterated the library; we will have expanded it into an even more welcoming and accessible space."[31]

Each library will take the adventurous yet often dangerous path it thinks will best meet the needs of the communities it serves. Some paths will be adventurous, even dangerous; others will not. And unlike in past generations, future libraries will resemble one another less and less. Our landscape will contain libraries that look and act as they do today, libraries that look similar to today's but that act differently, and libraries that bear little resemblance to today's–libraries that do not exist as physical places so much as they are integrated firmly into the fabric of their universities. Some will thrive; others will not. We should all strive to maximize the thrill of making valuable information available to more people with less effort. The libraries that do that will epitomize success.

NOTES

1. Beinhart, Larry. *The Librarian.* NY: Nation Books, 2004, 72.

2. *Silicon, Carbon, Culture: Combining Codes Through the Arts, Humanities, and Technology: Final Report of the Silicon, Carbon, Culture Initiative.* Urbana, IL: University of Illinois at Urbana-Champaign, 2004. www.las.uiuc.edu/scc/.

3. Lougee, Wendy Pradt. *Diffuse Libraries: Emergent Roles for the Research Library in the Digital Age.* Washington, DC: Council on Library and Information Resources, 2002.

4. Kaufman, Paula T. "How To Say 'No' and 'Why' Diplomatically." Kansas City, KS, Special Libraries Association, 1978.

5. Gillmor, Dan. "A Front-Row View of a Historic Time." *SiliconValley.com*, January 2, 2005. http://weblog.siliconvalley.com/column/dangillmor/.

6. Kurzweil, Ray. *The Age of Spiritual Machines: When Computers Exceed Human Intelligence.* NY: Viking, 1999.

7. Sullivan, Danny. "Google Scholar Offers Access to Academic Information." *SearchEngineWatch.com* November 18, 2004. http://searchenginewatch.com/searchday/article.php/3437471.

8. Price, Gary. "Google Partners With Oxford, Harvard & Others to Digitize Libraries." *SearchEngineWatch.com.* December 14, 2004. http://searchenginewatch.com/searchday/article.php/3447411.

9. "The Vanishing Mass Market." *Business Week*, July 12, 200:, 61-68.

10. "Channeling the Future." *Business Week*, July 12, 2004: 70-72.

11. Ibid.

12. Guskin, Alan E. and Marcy, Mary B. "Dealing with the Future Now: Principles for Creating a Vital Campus in a Climate of Restricted Resources." *Change*, 35:4 (July/August 2003):10.

13. Guthrie, Kevin. "An Introduction to Ithaka." *ARL Bimonthly Report* 236. Washington, DC: Association of Research Libraries (October 2004): 1-5. http://www.arl.org/newsltr/236/ithaka.html.

14. Lougee, Wendy Pradt Lougee. *Scholarly Communication & Libraries Unbound: The Opportunity of the Commons.* Workshop on Scholarly Communication as an Information Commons. Bloomington, IN, Indiana University, March 31-April 2, 2004.

15. Lynch, Clifford. "Digital Library Opportunities." *The Journal of Academic Librarianship* 29:5 (September 2003): 286-289.

16. Petrides, Lisa A. "Knowledge Management, Information Systems, and Organizations." *Research Bulletin.* Educause Center for Applied Research (ECAR). 2004:20, September 28, 2004.

17. Wittenberg, Kate. "Collaborators in Communication: Publishers, Scholars, and Information Technologists." *EDUCAUSEreview*, 39:6 (November/December 2004): 65-76.

18. Lougee, Wendy. *Scholarly Communications & Libraries Unbound: The Opportunity of the Commons. op. cit.*

19. *Ibid.*

20. Lynch, Clifford. *op. cit.*

21. Blossom, John. "Crystal Ball Redux: Looking Back on Shore's 2004 Forecast–and Peeking at 2005." *Commentary.* Shore, December 27, 2004. http://www.shore.com/commentary/newsanal/items/2004/2004127review.html.

22. Wilder, Stanley. "Information Literacy Makes All the Wrong Assumptions." *Chronicle Review*, (January 7, 2005): B13.

23. OCLC Online Library Computer Center. *The 2003 Environmental Scan: Pattern Recognition.* Dublin: Ohio, 2003: 78-81.

24. Kaufman, Paula T. "Whose Good Old Days Are These? A Dozen Predictions for the Digital Age." *Journal of Library Administration* 35:2, 2001.

25. See http://www.illinoisvlt.net/ as one example.

26. National Research Council (U.S.). Panel on the Impact of Information Technology on the Future of the Research University. *Preparing for the Revolution: Informa-*

tion Technology and the Future of the Research University. Washington, DC: National Academies Press, 2002.

27. "The Power of Design." *Business Week* May 17, 2004: 86-94.

28. Blossom, *op. cit.*

29. OCLC. op. cit.

30. Murray, Janet H. "The Exhilaration of Access." *Threshold*, Winter 2004.

31. *Ibid.*

doi:10.1300/J111v46n01_02

Google, Libraries, and Knowledge Management: From the Navajo to the National Security Agency

Dennis Dillon

SUMMARY. This paper is an exploration of taxonomy, discontinuity, culture, lost knowledge, planning the semantic Web, and how pre-network concepts such as libraries and knowledge management fit into a linked and searchable networked environment. It will examine how we capture, re-purpose, and make available what is important in a complicated world–a world in which our database can tell us both how to manufacture toothpaste and who is likely to buy it, but not what life forms exist on earth, or even what the chances are that your job will exist five years from now. doi:10.1300/J111v46n01_03 *[Article copies available for a fee from The Haworth Document Delivery Service: 1-800-HAWORTH. E-mail address: <docdelivery@haworthpress.com> Website: <http://www.HaworthPress.com> © 2007 by The Haworth Press, Inc. All rights reserved.]*

KEYWORDS. Knowledge management, tacit knowledge, libraries and digital information, librarians in the digital world

Dennis Dillon is Assistant Director, University of Texas at Austin Libraries, General Libraries–Administration Office, 1 University Station Stop S54000, Austin, TX 78712 (E-mail: dillon@mail.utexas.edu).

[Haworth co-indexing entry note]: "Google, Libraries. and Knowledge Management: From the Navajo to the National Security Agency." Dillon, Dennis. Co-published simultaneously in *Journal of Library Administration* (The Haworth Information Press, an imprint of The Haworth Press, Inc.) Vol. 46, No. 1, 2007. pp. 27-40; and: *Digital Information and Knowledge Management: New Opportunities for Research Libraries* (ed: Sul H. Lee) The Haworth Information Press, an imprint of The Haworth Press, Inc., 2007. pp. 27-40. Single or multiple copies of this article are available for a fee from The Haworth Document Delivery Service [1-800-HAWORTH. 9:00 a.m. - 5:00 p.m. (EST). E-mail address: docdelivery@haworthpress.com].

One of the most enduring and simplest of human institutions is the Japanese Shinto shrine, some of which have been built and rebuilt in the same location for over a thousand years.

SHINTO

When you approach a Shinto shrine you pass through gates guarded by a pair of stone lion dogs. The lion dogs were originally two different entities, a lion and a dog, and they were very different in appearance. Over the generations human stonecutters found it easier to carve them in the same proportions until their figures blended and they became one. Except that one lion dog has his mouth open representing the first breath you take when you are born, and the other has his mouth closed, representing the last breath you take when you die. Between the two lies all of existence and this is what you pass through when you enter the Shinto shrine.[1]

Shinto has little in common with modern information systems. Shinto has no sacred texts, no hierarchy, no body of knowledge–and as far as Westerners are concerned, it seems to have little or no information. It doesn't make any difference to Shinto if the Internet, or libraries, or knowledge management systems exist or not.

Right now, however, the Internet is giving every librarian pause. Recent innovations like Google Scholar, and Google's announcement that it will digitize millions of books, put us somewhat in the position of the local mom and pop store that has just learned that Wal-Mart will be moving in next door.

The usual response to the competition presented by Wal-Mart is for the smaller retailers to talk about the special services they can provide, how they will begin stocking niche products that Wal-Mart is not interested in, and how they will re-focus on providing quality and convenience. In other words, they plan to move upscale and sideways and concede the mass market to Wal-Mart. It's all very brave talk, and sometimes a few of them are still in business five years later.

GOOGLE

We are all aware that Google and similar Web tools have redefined the popular perception of the very nature of information. Google has remade the landscape in which our industry operates. It is always there at the exact moment of need. And in many cases, it provides information that is good enough or better than what the library can provide, and it does so within seconds without requiring advanced library degrees or special training. However, Google remains fundamentally different from the values and aims that libraries have historically celebrated.

One of the questions that librarians could fruitfully have tattooed on strategic parts of their anatomy is a simple one: "What game are we playing?" Are we playing the library game or are we playing the information game? I know how to play the library game. The library game has been refined over generations and we are all familiar with what is required to be a successful player. The same thing cannot be said of the information game. When I'm playing the information game I find myself asking questions about how I can be an effective publisher–is it best to open an information repository, host an open access journal, or participate in any one of an endless number of joint digital library projects? In the information game I worry about global competitors, like Google, whose aims and motivations I do not understand. In the information game I worry about things like information literacy and learning objects and federated searching, and about discontinuing or outsourcing traditional library functions in order to participate in still ill-defined new information opportunities. And all this is while knowing in my heart that though these things are rational short-term responses to current institutional conditions, they may be as irrelevant to our long-term survival strategy as bustles, beehive hairdos, and other fashions of the day. But holding onto the comforting concepts we tell ourselves on this morning's listservs and at tomorrow's conferences is what provides us with that much cherished flash of hope; somewhat like the momentary relief a drowning man feels when you throw him an anchor.

Of course few among us are so simple-minded as to believe that either technology or the Web is the answer. If it were, then individuals with the most technology would long ago have become the most fulfilled people on the planet. However, in polls taken every year since 1946, the happiest people in America have been the Amish, precisely those Americans with the least access to technology.[2] We all know in our soul that technology is only one of several solutions to the problems that beset mankind, but technology does have the power to change the question, and that is what it has done to libraries. As a result, some of the original reasons for our existence and the questions that we no longer ask our users may turn out to be the ones that count: Do libraries contribute to your sense of happiness and quality of life? Are libraries an important part of the community common good? Are libraries a place of comfort, optimism, and intellectual excitement? In a world filled with stridency and competing commercial messages, is the library still a place of integrity? Are libraries important to your children, parents, and friends, and does the library contribute to your job performance, personal fulfillment, and assist you with life problems? There are other questions we could ask.

People's motivations and desires are complex and I don't believe that even librarians understand the full range of values that libraries bring to their funding community. One of the conclusions of the recently completed Pew Internet and American Life Project was that the Internet would increasingly blur the boundary between work and leisure, thereby changing the nature of the home and family dynamics.[3] This is unfortunate on a multitude of levels. My wife and I always took our kids out into the wilderness for a month every summer, far away from the telephone and TV, just so we could all interact without the beeping distractions of technology. Very quaint, I know, but libraries may be part of just such a needed antidote to the modern commercialized life style. Libraries are flexible: we can be fast, we can be slow, we can be whatever the user needs. Libraries are one of the few places in our society where the mind, your mind, is celebrated, pampered, and treated with respect and appreciation.

On a more practical level, knowing what game we are playing does affect how we allocate our resources and plan for the future. Should we buy that new integrated library system, or wait for all library resources to be available through Google? Should we reduce reference and cataloging staff and move them into new jobs supporting institutional repositories and hosting open access journals and learning objects? Does it make sense to promote face-to-face interactions through the reference desk, or should those resources be shifted toward developing a simpler and more intuitive Web presence? Is the library more competitive if it provides the highest possible quality cataloging and reference assistance, or if concentrates on efficiency and throughput, while devoting its precious discretional energy to the new but unproven niche areas like institutional repositories? All of these are questions that cut quick to the soul of what it means to be a librarian. Of course in the back of our minds, we all know that the ultimate test of our effectiveness will be how well we meet the needs of our funding communities. Last December Salinas, California became the first city in the country to shut down its library system; others have been saved by last second budgetary compromises. Consultants in the UK recently recommended that various public service librarians at Bangor University be fired because as the consultants say "it is hard to justify the value-for-money terms at a time when the process of literature searches is substantially deskilled by online resources."[4] In other words, library funding is in danger unless we can make a demonstrated and appreciated difference in our communities.

One of the problems libraries face is modernity itself. We were created to serve the needs of a local community, and yet our competitors are increasingly global, as are the interests of our users. Libraries feel the effects of globalization through Google. And in some senses Google is the face of modernity with its focus on choice, individualism, and instant gratification, all partaken of without context, or an enduring foundation, or a stable value system. Google returns truth and falsehoods in the same result set, and presents them all without context–in a sense plucking information particles out of the eye of a whirling and ever changing information hurricane based on algorithms that are only slightly more sophisticated than popularity contests, perfect for a generation that grew up flipping between a hundred TV channels with a remote glued to one hand and a video game controller in the other.

As human beings of the early twenty-first century, this information environment is part of the daily warp and woof of our turbulent lives. As librarians, we must choose how to adapt our mission to fit these new realities. Part of what makes being a librarian so interesting today, is that we are in the midst of trying on a new identity in what is suddenly a much more networked and inter-connected world. But as one of the first moderns, Oscar Wilde, cautioned, "It is only the modern that ever becomes old fashioned."[5] Meaning, of course, that enduring values are enduring for a reason.

NAVAJO

By some accident of wandering fate over the last twenty years, I've frequently found myself driving through long southwestern nights listening to KTNN, the voice of the Navajo nation–AM radio out of Window Rock, Arizona. The Navajo have much more critical identity questions than those that face librarians. Caught between the frivolous white man and an endangered traditional culture–the Navajo are struggling to find the best way forward, yet 72% of Navajo households do not even have a telephone. Because of this, AM radio has been an effective way to pass on needed information and instill tribal cohesion. Yet, long before the advent of radio, the Navajo were able to transmit the values and stories of their culture from generation to generation through oral tradition. Today they possess an unusually powerful belief system that has allowed them to thrive while other Indian cultures withered. Recognizing the power of oral traditions is one of the founding principles of knowledge management.

KNOWLEDGE MANAGEMENT

The architects of knowledge management systems talk about two types of knowledge. One is explicit knowledge. This is the type of knowledge that resides within libraries. It is articulated, collected, codified, and written down. The other type of knowledge is tacit. Tacit knowledge is personal knowledge. It is what is being shared when experts show the ropes to novices. It is the knowledge passed on in the mentoring relationship. It is what parents despair of ever teaching their children. It is what thoughtful teachers are able to convey to their students. This type of knowledge is difficult to acquire and even more difficult to pass on, but it is very sticky because it is communicated personally. As Hemingway noted, "There are some things which cannot be learned quickly—and time, which is all we have, must be paid heavily for acquiring."[6] One of the things Hemingway was talking about is tacit knowledge. People may forget the explicit knowledge they read in books, but they seldom forget a critical bit of wisdom that has been nailed into their heads by the hammer of life, or passed on by their mother.

Most knowledge management research revolves around efforts to capture those elements of tacit knowledge that are important to the organization, and then to make this knowledge useful to others. The tacit knowledge that an organization's employees possess is typically the most valuable asset that an organization has. It is tacit knowledge that is needed during times of change or emergencies. And it is tacit knowledge that separates successful organizations from unsuccessful ones.

AVANTIS

Avantis is the largest pharmaceutical company in Europe. Its employees use an e-mail system called Tacit K-mail that watches all of the company's e-mails and creates a searchable database of the daily communications of its far-flung research scientists. Avantis staff can then search through a database of these e-mails, looking for nuggets of tacit knowledge that will help the company to research and develop new drugs. The explicit information available in journals and databases is also helpful, but it is difficult and time-consuming to use. By using outside consultants who are already familiar with the published literature, and combining this with their internal knowledge management system, Avantis has been able to become more efficient and productive. In essence, Avantis is availing itself both of the tacit knowledge of their consultants who understand the literature, and the tacit knowledge of their own research scientists.

NSA

Universities could similarly capture the tacit knowledge residing in their faculty by recording their e-mails, phone conversations, and lectures, and putting this information in a central searchable database, but then again, the National Security Agency (NSA) has already done some of this work for us.

Stored within its eleven acres of supercomputers in Ft. Meade, Maryland, and at hundreds of other sites around the world, the NSA uses a system known as Echelon to monitor the world's e-mail, phone conversations, faxes, satellite communications, and the like. Analysts operating out of standard corporate cubicles listen to recordings of intercepted phone conversations, while looking at the photographs of their subjects and examining their various communications.

A standard NSA office will process a million messages in thirty minutes. Using a software filter known as "dictionary" these million messages will be reduced to one thousand messages that are forwarded to the analysts, from which ten are then selected for special attention, with the result being one report that will then be entered into the knowledge management database.

An authorized user can then go to the NSA Web site and find the intercepts organized by categories. Under each category are listed the subject lines of the reports. When you open up the analyst's report, relevant e-mails, faxes, and other information are attached.

NSA's activities essentially encompass all of the world's private communications and are much larger in scope than most knowledge management efforts. In some ways they represent a cross-fertilization of both standard data warehousing and data mining techniques along with cutting edge developments in knowledge management. But at its heart NSA's concern is with capturing and making sense out of tacit information in order to adapt counter-terrorism efforts to today's rapidly changing and discontinuous environment–or to state it another way, they are employing knowledge management techniques in precisely the way that they are designed to work.

However, like all knowledge management systems, Echelon is not foolproof. As one official said, "Supposing you pipe every communication that goes on in China back to the United States. Then you've got to have somebody process it; you've got to have a linguist listen to it. And the chances of your ever digging out from under the pile of information and finding what's important are miniscule."[7] This problem of being able to effectively sort through the world's information in order to find what is critical haunts all of us in the information industry, as does the related worry that we might not be collecting the right stuff.

KOJIMA MONKEYS

A few years ago the monkeys on Kojima Island made international news because the females were teaching the younger monkeys to wash sand off their food before they ate it. Supposedly this is something that only humans should know how to do. The monkeys on the nearby island of Torishima did not do this. Then one day the monkeys of Torishima also started washing their food. The Japanese academics that had studied these monkeys for years were mystified, and all sorts of fantastical academic theories were suggested to explain the occurrence. "Was there a recessive food washing gene? Could the monkeys of Torishima have peered across the water and only now understood what the other monkeys were doing and copied them? Could this be some sort of simian extrasensory perception?" Then someone asked a local fisherman about this behavior and he said–"well the monkeys swim back and forth between the islands all the time–maybe that has something to do with it." And that was how the academics discovered the swimming monkeys of Kojima. This was common tacit knowledge among the locals, but a complete mystery to the academics. This is a classic problem for all information systems. The collecting line always has to be drawn somewhere, and whatever is not in the collection will at some point reveal the system's limits.[8]

As librarians, we are well aware that no information tools are culturally neutral. The library is rife with well-developed philosophies and values that both help and hinder our response to the changing environment around us. We are not neutral in the way we approach the world of information, and neither is Google or knowledge management. Each information tool embodies a different set of deeply ingrained information prejudices, preconceptions, and beliefs about the nature of the world and about how people use information. Each tool starts from different underlying principles and they each produce different results.

Knowledge management, libraries, and Google do have some things in common; all three are ways of organizing portions of reality so that each of us will be better equipped to deal with the challenges that the universe and our daily lives throw at us.

Google organizes its version of reality by omnivorously collecting digital content. It does not collect analog formats. It does not collect and provide access to private information such as e-mails, faxes, diaries, and phone conversations. It does not collect and provide access to licensed information such as e-journals and e-books. It does not validate or provide context for the information it makes available. Unlike libraries, it retrieves unfiltered information particles. These particles may be a book, a picture, a teenager's unintelligible Web page, an advertisement, or even blatant propaganda, and all of these information particles are then presented without context.

Google is the same for everyone. It is not tailored for different user groups, and it does not change, as local user needs shift. If you have a problem, there is no Mr. Google to talk to. Because it is constructed to serve every breathing individual on the planet today, and all from one interface–both kindergarteners and Nobel Prize winners who input the same search string, will retrieve the same search results. Its very nature is different from that of libraries.

SCORPION AND FROG

I have never met anyone who believes that Google Incorporated has evil intent, but it is not Google per se, that libraries should be concerned about. It is the alluring combination of search technology and that eternal temptress, capitalism, which should give us pause. The essential nature of Google is different from the essential nature of libraries. There is an old parable that may help illustrate why this matters.

A scorpion and a frog meet on the bank of a stream and the scorpion asks the frog to carry him across on its back. The frog asks, "How do I know you won't sting me?" The scorpion says, "Because if I do, I will die too."

The frog is satisfied, and they set out, but in midstream, the scorpion stings the frog. The frog feels the onset of the paralysis and starts to sink, knowing they both will drown, but he has just enough time to gasp "Why?"

The scorpion replies, "Because it is my nature."

In this parable the stinging scorpion may be the simple advance of technology, it may be cost cutting administrators who don't see why they need a library because everything will soon be on the Web, it may be the innate impersonal forces of capitalism, it may be unreflecting users who simply vote with their feet, or it may be a combination of people who are simply playing a different game with different values than we are. It doesn't really matter. We could wake up tomorrow to the news that Halliburton has purchased Google and intends to turn it into a private corporate information tool, or that a Hong Kong bank has now seized controlling interest and is converting Google's content into Chinese. This is the nature of people and organizations that are not playing the same game as libraries.

One of the common dilemmas of all information tools is one that has been given centrality in the theories of knowledge management. This problem is encapsulated in a classic 1971 quote from Charles West Churchman, who said, "Knowledge resides in the user and not in the collection. It is how the user reacts to a collection of information that matters."[9] Obviously, this is just as true of libraries and Google as it is of knowledge management. One thing Churchman's observation has done is to subtly shift the locus for measuring the effectiveness of any information tool, to the effectiveness of the match between the user and the information the user retrieves.

In other words, we once again come back to the age old human problem best expressed by Dorothy Parker's observation: "You can lead a whore to culture, but you can't make her think."[10] A variation on this problem occurred in the 1960s when NASA decided to test some high altitude cameras by taking photos of Yellowstone. When Bob Christiansen of the U.S. Geological Survey saw them he was shocked. He had been looking for the Park's volcano, which he knew had to be there somewhere, but what he saw in the photos was that virtually all 2.2 million acres of the park was a volcano. It was in fact a super volcano and the last explosion had left a crater more than 40 miles across–what we now call Yellowstone Park.[11]

This is one of those cases where no one had been able to see the forest, because all the trees were getting in the way. It took Bob Christiansen's Eureka moment, what neuroscientists call "insight," to make what now seems like an obvious connection. In this case there was an effective match between the collection and the user, along with the addition of insight.

DIFFERENCES

Libraries, knowledge management, and Google cannot always be equally effective in their matchmaking capabilities. They differ in what they collect and provide access to. They differ in formats, in their relation to copyright and privacy, and in the validation they give to the information they select. They differ in the audiences served. They differ in the levels of context they supply for the information in their purview. They differ in community. They differ in breadth. They differ in collectible units. They differ in their approaches to organization and classification of material. They differ in the amount of human oversight and human responsiveness they bring to the collection and to user concerns. They differ in how tailored both the supporting organization and the collection is to community and individual needs. They differ in the amount of additional services that are provided to users. They differ in terms of place. They differ in their commitment to carrying the collection forward over time. And they differ in their commitment to promoting the underlying values of civilization that have come to define humanity.

All three of these information tools can be effective when they are employed by the right hands for the right purposes. They can all assist in promoting the original and innovative thinking that results in creativity and insight. However, these are actions that occur in the brains of individuals. And these individuals are all members of different cultures with different conceptions as to the nature of knowledge, different ways of approaching problems, and who will interpret the results of these information tools in different ways. As Churchman said, it is how the user reacts to the collection that matters.

Americans, for example, celebrate individualism above all other values, and we confidently believe that we can plan and control the future through pragmatically dividing every problem into ever smaller discrete units. If you look at world history, this is a very odd way of approaching both the nature of knowledge and the scope of human problems. But it is our nature, and we ask our information tools to work within these pre-conceptions.

On the other hand, the British strive to manage situations, while believing it is arrogant to assume you can control the universe and everything in it. The French value rationality and strategic planning while recognizing and celebrating the sensual and irrational side of mankind–a duality the rest of the world sometimes finds baffling. The Chinese believe in patience and harmony, and that knowledge arises in an indefinite and ever changing process that only simpletons would attempt to reduce to grids and metadata. The entire Japanese system of logic is based on maintaining harmonious relationships, and as a result they value indirectness, taciturnity, purposeful ambiguity, and evasiveness in communication–historically making them difficult customers for knowledge management systems that attempt to ferret out the heart of tacit communication. While Indians (and other Buddhist influenced cultures) are very skeptical of attempts to reduce knowledge to language or taxonomies, and view attempts at categorization and measurement as purposefully deceitful and actively misleading.

Just hinting at some of these different underlying cultural views of the nature of knowledge reveals the complexity of the problems that underlie developing effective knowledge handling tools that will be equally effective for all cultures and types of information. The current goals of Google, to identify, particularize, and make freely available as much of the world's digital information as possible, are very different from library values in which explicit information is carefully selected and placed in context for a specific audience, and the values of knowledge management which attempts to unveil tacit knowledge connections among workers in very specific fields.

END OF LIBRARIES

On a more mundane level, Google nevertheless, has the power to accidentally remove libraries from the face of the earth. I'd be sorry to see this happen, but if Google remains free, and if it contains all the books ever written, and indexes every article ever produced on earth–no matter how much people may value libraries, there are going to be those who question whether the role that libraries play in society is worth the cost. No one can predict how the tens of thousands of administrators in charge of our funding will react to these changes. Defining our value and our role in crystal clear fashion, and finding an appropriate place in this information universe is our challenge.

One experience that every librarian has had in recent years is to watch his or her libraries empty out when the local connection to the Internet goes down. Suddenly everybody leaves the building, even though the physical books and journals that drew them to the library are still there.

If we remember the history of Shinto, then we may find comfort in this behavior. If people are leaving the library when the Internet goes down, then the cold hard facts are that whatever we may imagine that our libraries are, what we are in reality is a supplement to the Internet. We may have delusions that this isn't so, but those fantasies shrivel every time new content is added to the Web, and every time a child sits down to discover the world that exists inside a computer screen.

As you recall, Shinto shrines originally were guarded by two different beasts, a lion and a dog, but over time these beasts blended together and became one–just as the Internet and libraries are melding together to form part of the continuing spectrum of information services. In Japan, Shinto continues to exist as a belief system alongside late arriving, new, and competing explanations of the mysteries of life in the form of Buddhism and Confucianism. The Japanese have not discarded Shinto, instead they have amalgamated the strengths of each of these different sets of beliefs into their daily life–just as we in the West have adapted to changing technology and still use postal mail services, even though the telegraph, telephones, faxes, and e-mail all perform a similar function; and just as we still listen to radio despite the invention of television, MP3 players, and the Internet.

Herein lies our future. It lies in how the user reacts to the collection of information services we provide. It lies in improving our user's productivity, creativity, innovation, personal fulfillment, and yes, even happiness. It lies in more forcefully defining ourselves as a place of the mind where individuals are respected and treasured and made more productive and fulfilled. No organization is going to fund us on the flimsy premise that librarians are better equipped to search and interpret Google results, likewise there is not likely to be much future in digitizing things when the vast majority of information is born digital. Like the scientists studying the Kojima monkeys or searching for the Yellowstone volcano, libraries have become so enamored of technology that we sometimes cannot see what is in front of our faces, which is that there are still people in our buildings and they are there for a reason. To carve an effective and sustaining pathway to the future, we have to be honest about what our strengths are and aren't, about the entire complex of things that today's information seekers value, and to acknowledge that though the history of technology lurches forward in uncertain steps, it generally serves to make it easier for the average human being of one generation to perform the tasks that required special training in the previous generation.

What this means is that users will increasingly be able to get the information they need on their own, and that the future of libraries is likely to be where it has always been, in providing a particular community with value-added information services that they cannot get elsewhere. As freely available information tools make life easier for the average user, the niche that librarians will inhabit will increase in sophistication. Our future marketing slogan may very well be a variation of, "when the Internet isn't good enough," and the service we provide could be characterized as "Internet Plus"–meaning we provide access to information that is not easily available over the Internet, and to information in context that makes a difference, as well as a place where information seekers can gather. This is a comfortable and distinctive non-profit role for libraries that separates us from the globalized and profit-minded Google, and that celebrates what lies at the heart of our value system, which is taking care to connect users with the information they need for the betterment of us all.

In essence, our role has always been the collection and redistribution of the content that defines what it means to be human. No matter what the technological developments of our age, our mission is ultimately to collect and provide the best record of our species–its foibles, suspicions, hopes, wishes, and knowledge–that is in our power to do so. So when you go home tonight, consider the best of what humanity accomplished today–and make sure that is in your library.

NOTES

1. Ferguson, Will. Hokkaido Highway Blues: Hitchhiking Japan. (New York, Soho Press, 1998) p. 30.
2. Surowiecki, James. "Technology and Happiness: why more gadgets don't necessarily increase our well being," *Technology Review*, January 2005 pp. 72-76.
3. The Future of the Internet. (Pew Internet and American Life Project. Washington D.C., 2005) p. iii.
4. The Columbia Dictionary of Quotations ed. by Robert Andrews. (New York, Columbia University Press, 1993) p. 596.
5. Curtis, Polly. "Bangor Librarians Face Internet Threat" The Guardian Unlimited, February 16, 2005. http://www.guardian.co.uk/uk_news/story/0,3604,1415830,00.html.
6. The Columbia Dictionary of Quotations ed. by Robert Andrews. (New York, Columbia University Press, 1993) p. 513.
7. Bamford, James. Body of Secrets. (New York, Doubleday, 2001) p. 459.
8. Ferguson, Will. Hokkaido Highway Blues: Hitchhiking Japan. (New York, Soho Press, 1998) pp. 41-42.
9. Churchman, C.W. The Design of Inquiring Systems, Basic Books. (New York, NY, 1971) pp. 9-11.
10. www.kirjasto.sci.fi/dparker.htm.
11. Bryson, Bill. A Short History of Nearly Everything. (New York, Broadway Books, 2003) pp. 224-235.

doi:10.1300/J111v46n01_03

The Scholarly Work of Digital Libraries

Judith M. Panitch
Sarah Michalak

SUMMARY. In the past, librarians have assembled and interpreted collections for smaller audiences that frequently were limited to one-on-one consultations in a highly specialized environment. Digital technologies, however, make it possible to combine, analyze, interpret, and disseminate the content of our collections in ways never before possible. This paper will compare and contrast two digital library programs and their innovative approaches to scholarship and knowledge creation and dissemination. doi:10.1300/J111v46n01_04 *[Article copies available for a fee from The Haworth Document Delivery Service: 1-800-HAWORTH. E-mail address: <docdelivery@haworthpress.com> Website: <http://www.HaworthPress.com> © 2007 by The Haworth Press, Inc. All rights reserved.]*

KEYWORDS. Scholarly values and digital libraries, digital projects in academic libraries, digital libraries

When Sarah moved last year from the University of Utah's Marriot Library to the University of North Carolina at Chapel Hill, she immedi-

Judith M. Panitch is Director of Library Communications, University of North Carolina at Chapel Hill, CB# 3900, Chapel Hill, NC 27514-8890 (E-mail: panitch@email.unc.edu).

Sarah Michalak is University Librarian and Associate Provost for University Libraries, University of North Carolina at Chapel Hill, CB# 3907, Davis Library, Chapel Hill, NC 27514-8890 (E-mail: smichala@email.unc.edu).

[Haworth co-indexing entry note]: "The Scholarly Work of Digital Libraries." Panitch, Judith M., and Sarah Michalak. Co-published simultaneously in *Journal of Library Administration* (The Haworth Information Press, an imprint of The Haworth Press, Inc.) Vol. 46, No. 1, 2007, pp. 41-64; and: *Digital Information and Knowledge Management: New Opportunities for Research Libraries* (ed: Sul H. Lee) The Haworth Information Press, an imprint of The Haworth Press, Inc., 2007, pp. 41-64. Single or multiple copies of this article are available for a fee from The Haworth Document Delivery Service [1-800-HAWORTH, 9:00 a.m. - 5:00 p.m. (EST). E-mail address: docdelivery@haworthpress.com].

Available online at http://jla.haworthpress.com
© 2007 by The Haworth Press, Inc. All rights reserved.
doi:10.1300/J111v46n01_04

41

ately recognized the central role that the digitization programs of both libraries play in current priorities and planning. Just as striking were significant differences between the two programs in terms of their scope, procedures, and online features. More subtly, it seemed that North Carolina's digital library expressed in its rhetoric, particularly through frequent comparisons with publishing, a clear consciousness of being "scholarly," while this terminology has not generally been used at Utah. Wondering whether the Marriott digital library also had scholarly aspects, Sarah began to see that despite significant differences, both the Carolina and Utah projects adhere to some core principles that, while not always overtly articulated, suggest a shared rigor that could indeed be called scholarly. This paper began as an exploration of these shared scholarly values and procedures.

As we proceeded, we found ourselves thinking more critically not just about the projects at hand but also about what it means to produce a "scholarly" project or to be engaged in a "scholarly" enterprise. This is not an idle semantic exercise. Being scholarly is the yardstick for the academic environment in which we function. Wendy Pradt Lougee finds the research library becoming "more deeply engaged in the fundamental mission of the academic institution–i.e., the creation and dissemination of knowledge–in ways that represent the library's contributions more broadly and that intertwine the library with other stakeholders in these activities."[1] In building digital libraries, we engage in a complex set of activities through which we produce knowledge by selecting and presenting new combinations of primary resources together on the Web, and we create new opportunities for researchers and students to derive knowledge from these resources. Digital libraries are an ideal vehicle for the library to step into the role described by Lougee and to lead within our institutions. If we are serious about taking on the task of producing new knowledge in an academic context and wish to do it well, then we need to be cognizant of the rules of that game.

This paper, guided by recent thought in higher education about the definition of scholarly activity, will explore what is meant by a scholarly orientation and the ways these values find expression in the digital libraries at North Carolina and Utah. We realize that, whatever their merits and lessons, two projects represent only a small portion of the universe of scholarly digital library projects. Still, by studying them, we hope to illuminate how digital libraries contribute to the work of the university and to derive some guiding principles that might assist in evaluating future digital library enterprises. This paper presents not a conclusion but a starting point based on an admittedly small but, we hope, revealing sample.

WHAT DO WE MEAN BY "DIGITAL LIBRARY"?

In talking about the "scholarly work of digital libraries," two definitions are required. We will explore later the question of what is "scholarly," but it is equally fundamental to clarify what we mean by "digital library," a term that has taken on multiple values in recent years. For this paper, we use "digital library" in a fairly narrow sense to refer to collections of analog materials largely, although not exclusively, from the library's own holdings, that have been converted to digital form and made available online, plus the technologies and services that support those collections.

This definition is neither the most inclusive nor the most common. The Digital Library Federation, for example, has crafted a working definition with extremely broad applicability: "Digital libraries are organizations that provide the resources, including the specialized staff, to select, structure, offer intellectual access to, interpret, distribute, preserve the integrity of, and ensure the persistence over time of collections of digital works so that they are readily and economically available for use by a defined community or set of communities."[2] It is little wonder that a 2002 environmental survey finds considerable ambiguity surrounding the term "digital library": "even in the professional environment, a large number of definitions are in use."[3] Most current usage includes not only converted materials, but those "born digital" and acquired from any source, including products licensed and purchased from vendors, along with the underlying technical infrastructure needed to make them available, and tools such as the online catalog.

Why have we chosen a more circumscribed approach? In the first place, digital conversion is an activity with which many, many libraries have experience. An informal review of the Web sites of the thirty-two largest ARL libraries found that twenty-nine of them presented digital document surrogates assembled as coherent collections.[4] Yet, as Ann Okerson observes in a reflection on "The Future of Digital Libraries": "Great digital collections already exist, although I would submit that apart from the best-publicized ones, we don't know what many of these are, where they are or very much about them."[5] Perhaps less visibility has contributed along the way to less reflection. It does seem that, in the understandable need to grapple with pressing issues surrounding digital libraries broadly construed, the complexity and value of conversion and of providing the access and services associated with converted documents has been underestimated. Daniel Greenstein and Suzanne Thorin, for example, consider an emphasis on digital conversion to be characteristic of the "young" (as opposed to the "maturing" or "adult") digital library and speak of most early online collections with some skepticism as principally "technical experiments rather than as means of redefining scholarship." Most of these collections, they continue, were "too small to support more than very casual kinds of browsing[;] too idiosyncratic to be integrated meaningfully into larger virtual collections[; and] too passive to maintain a user's interest for very long."[6]

We do not believe that conversion activities ought to be so rapidly dismissed. Through our own experience with the two projects discussed in this paper, we have found that they certainly have the capacity to be deep, complex, sustained, transformative in their own right, and the source of important lessons for the many libraries grappling with the digital representation of analog collections.

Documenting the American South
University of North Carolina at Chapel Hill
(http://docsouth.unc.edu)

Documenting the American South (or DocSouth) began with the experimental digitization of a few highly circulating slave narratives that were made available online to the public in 1996. Those first texts became the core of a Web site that today consists of seven discrete collections intended to illuminate southern history, literature, and culture. The collections are: "First-Person Narratives of the American South," "Library of Southern Literature," "North American Slave Narratives," "The Southern Homefront, 1861-1865," "The Church in the Southern Black Community," "The North Carolina Experience, Beginnings to 1940," and "North Carolinians and the Great War." As of December 2004, the collections totaled 1,287 titles, with an additional 90 in various stages of production. DocSouth has also grown to include some 12,000 images–covers, title pages, illustrations, and images from the digitized texts, as well as graphic items such as photographs, artifacts, and posters. There are 25 audio files of oral histories and songs and a varied collection of supporting materials including biographies, summaries, essays, and contemporary news and magazine articles. By late

2004, the site was receiving over 4 million visits per month, which administrators conservatively translate to some 335,000 viewings of the primary texts that constitute the intellectual core of the site. An editorial board provides guidance and faculty editors are associated with nearly all of the collections. The site is managed by the library and hosted by ibiblio (http://www.ibiblio.org), formerly UNC's Sunsite, with which the library forged a productive cooperative relationship in the very early days of DocSouth.

Marriott Library Digital Technologies Projects
University of Utah
(http://www.lib.utah.edu/digital/)

The Marriott digital library builds on a tradition in the library of trying and adopting new technologies. With thirty distinct collections, the Marriott digital library today emphasizes the history of Utah and the Intermountain West, but is not restricted to this focus. Other holdings, some selected by the library and others digitized at the request of faculty members for classroom instruction or research projects, are highly varied and represent a variety of fields. Digital collections include Aztec Codices, the Dard Hunter Book Collection, Ethiopian Manuscripts, the Frank Lloyd Wright *Wasmuth Portfolio*, and online versions of scholarly journals and monographs published by the university and its press.

We call attention to three features. The first is an engagement with difficult formats such as maps and folios. The second is the Marriott digital library's leadership in bringing together collections from multiple institutions, including smaller libraries and historical societies that would not otherwise have the resources for a digital library program. The Mountain West Digital Library, which includes texts, photographs, and documents about the history and settlement of the region, and the Waters of the West Digital Library are examples. Finally, its greatest topical strength is the Utah Newspaper Project, inspired by a visit to the state by Nicholson Baker, author of *Double Fold: Libraries and the Assault on Paper*. Although the Marriott Library's papers had been microfilmed and discarded, back runs of rural weekly papers were languishing in various locations around the state. Library staff began to digitize rural weeklies from these collections, eventually developing an effective large-scale, high-volume process capable of converting daily papers, as well as weeklies. There are now 350,000 pages digitized or in process and the program continues to expand with the aid of IMLS and LSTA grants.

WHAT DO WE MEAN BY "SCHOLARLY"?
DIGITAL LIBRARIES AND SCHOLARSHIP RECONSIDERED

Difficult as it is to define the "digital library," defining "scholarly" has proven equally problematic. Most of us have an intuitive grasp of what is scholarly and what isn't. Our libraries certainly produce endless handouts, Web pages, and instructional materials to help students distinguish between "scholarly" and "popular" resources–usually on the basis of such criteria as reproducible research, peer review, and appropriate citations–but this approach does not really help us think about the unique ways that digital libraries create and contribute to knowledge. We know that our digital library projects can be used by scholars, although this cannot be the defining criterion since we also know that these online resources enjoy even greater use among "non-scholarly" populations. Interestingly, we have found the most vibrant and clearly articulated discussions of what is "scholarly" to have emerged from the national dialogue of the last decade and a half concerning faculty roles and responsibilities.

In turning to the literature of higher education, we seek to situate the library and the digital library clearly within the realm of inquiry and the academy. We are not attempting to equate all of what we do with all of what faculty members do, nor are we proposing an argument regarding faculty status for librarians.[7] We do believe that this discourse provides an excellent tool for broadening our perspectives and a framework for analyzing the inherent complexity and sophistication of activities that are too easy to think of as "just a service."

In 1990, as part of a sustained effort by the Carnegie Foundation for the Advancement of Teaching to examine and recommend improvements to the American higher education system, Ernest Boyer published his influential *Scholarship Reconsidered: Priorities of the Professoriate*.[8] Following the evolution of the American university from its colonial origins, Boyer determines that conceptions of scholarship have failed to keep pace with the real demands of the current environment. The traditional triad of "research, teaching, and service," he asserts, no longer provides an adequate way to evaluate the performance of faculty members:

> We believe the time has come to move beyond the tired old "teaching versus research" debate and give the familiar and honorable term "scholarship" a broader, more capacious meaning, one that brings legitimacy to the full scope of academic work. Surely,

scholarship means engaging in original research. But the work of the scholar also means stepping back from one's investigation, looking for connections, building bridges between theory and practice, and communicating one's knowledge effectively to students. (Boyer, 16)

At the core of Boyer's re-evaluation is the articulation of four distinct but related categories of scholarship: "discovery," or those activities closest to traditional pure research; "integration," which involves "making connections across the disciplines," as through multidisciplinary or interdisciplinary studies, and also "illuminating data in a revealing way, often educating nonspecialists, too"; "application," or the use of knowledge to solve real-world problems; and "teaching," an activity that lies at the origins of scholarship and that Boyer actively sought to revalorize (Boyer, 15-25).

Boyer emphasizes that no scholar–and no scholarly project–excels in all of these areas and, indeed, we have found that our digital libraries are most like "scholarship of application," although they exhibit elements of the other categories. We may think about pure *discovery*, for example, when a digital library project brings new knowledge to light. DocSouth's slave narratives segment has surfaced several previously unknown titles, while the "Library of Southern Literature" is built around a bibliography developed by a faculty member specifically for the project. We may think about *integration* in the way that both projects are broadly multidisciplinary. The Marriott Library realizes another approach to integration by bringing together holdings of multiple research institutions in a single searchable interface, with initiatives such as the "Mountain West Digital Library" and the "Western Waters Digital Library." The tremendous use made of both projects by those outside the academy–K-12 students and teachers, genealogists, journalists, independent researchers, and the curious public–speaks to Boyer's emphasis on "educating nonspecialists." While neither project was designed as a *teaching* tool, we have also become aware of their tremendous potential in both formal classroom teaching and self-directed learning. Both initiatives make a considerable effort to provide background information and opportunities for the user to learn more. DocSouth has begun systematically associating "Learn More" links containing summaries, biographies, and related materials with titles in the collections. DocSouth also features a new K-12 "Classroom" section (http://docsouth.unc.edu/classroom/) with a "Teacher's Toolkit" and sample lesson plans.

It is perhaps as *application* endeavors that the scholarship of the projects under study is most fully realized. A legacy of the nineteenth- and early twentieth-century emphasis in higher education on practicality and engagement, the scholarship of application seeks to use knowledge toward the resolution of "consequential" problems. What makes these activities scholarly, observes Boyer, is their dynamic nature; the outcome is not simply the development of practical solutions, but the achievement of new discovery and deeper insight through the process: "New intellectual understanding can arise out of the very act of application–whether in medical diagnosis, serving clients in psychotherapy, shaping public policy, creating an architectural design, or working with the public schools. In activities such as these, theory and practice vitally interact, and one renews the other" (Boyer, 23).

As librarians, we are comfortable in the role of application scholars, even if we do not always think of our work in this way. We bring our expertise to bear constantly in the challenges of selecting and acquiring resources, making them available, connecting our users to the items they need, providing interpretation, and preserving materials for the long-term. In this sense, "digital libraries can be viewed as extensions and augmentation of traditional libraries" (Allard, 234).

To this foundation, digital libraries also add new requirements regarding expertise in the capacity of technologies. Technology is thus both the impetus for asking those new questions of application–How can we use emerging technologies to bring selected library collections and services to the online environment? How can these technologies be used to enhance the experience of using the library and its resources?–and the means of realizing their answers. Clearly, this is not simply an instance of rote or perfunctory implementation of existing principles and practices. We will discuss these contributions in greater detail later, but it is not hard to see how attempting what has not been done before will inevitably lead to novelty, to invention, and to new understanding, true to the spirit of scholarship.

Boyer's inclusion of architecture in his brief listing of "scholarly application" activities seems particularly apt. Just as architects draw on many types of knowledge–physics, geometry, design, engineering, planning, and so on–to envision and realize an entirely new construction out of raw materials and space, so our digital library activities take the "raw materials" of our collections, along with the accumulated knowledge of our own and other fields, to create substance and structure where it was not. We not only produce new scholarly output ourselves and enable inquiry by others, but we add to the knowledge base of our institutions and our field in the process.

ASSESSING THE SCHOLARLY VALUE
OF THE DIGITAL LIBRARY

It is not enough to say that digital libraries by their nature and goals are compatible with a broadened view of scholarship and therefore are scholarly. Indeed, one of the common critiques of Boyer's study is that it fails to provide benchmarks by which scholarship other than traditional "discovery" can be evaluated. If we are to claim our place as part of the scholarly activities of the university, we must ask how well we do what we are doing. Does our activity rise to the level of scholarship and, if so, in what ways?

The challenge of developing criteria for evaluation of scholarship was taken up, again by the Carnegie Foundation for the Advancement of Teaching, in a 1997 study, *Scholarship Assessed: Evaluation of the Professoriate.*[9] Working from a variety of sources–including guidelines on hiring, tenure, and promotion at colleges and universities across the country; information from granting agencies, editors, and directors of scholarly journals and university presses about standards in use to decide the scholarly merit of proposals and manuscripts; and student and faculty peer evaluations for teaching–the authors were able to discern characteristics shared by all four types of scholarship. "We have found," they write, "that when people praise a work of scholarship, they usually mean that the project in question shows that it has been guided by these qualitative standards: (1) Clear goals; (2) Adequate preparation; (3) Appropriate methods; (4) Significant results; (5) Effective presentation; (6) Reflective critique" (Glassick, Huber, and Maeroff, 24-25). Using these characteristics, are there, then, more specific ways that we can think about the "scholarly work of digital libraries"?[10]

Clear Goals and Adequate Preparation

In thinking about goals, the authors of *Scholarship Assessed* recommend asking such questions as: "Does the scholar state the basic purposes of his or her work clearly? Does the scholar define objectives that are realistic and achievable? Does the scholar identify important questions in the field?" (Glassick, Huber, and Maeroff, 25). In assessing the level of preparation, questions to ask include: "Does the scholar show an understanding of existing scholarship in the field? Does the scholar bring the necessary skills to his or her work? Does the scholar bring together the resources necessary to move the project forward?"(Glassick, Huber, and Maeroff, 27). We combine these two elements here because they are intertwined in the origins of the projects.

Documenting the American South was envisioned from the beginning very much as a collection development project driven by a specific body of content. Early discussions held in the library in 1991 and 1992 eventually resulted in an internal memo drafted by Pat Dominguez (Humanities Bibliographer), Luke Swindler (Social Sciences Bibliographer), and David Moltke-Hansen (Director of the Southern Historical Collection) proposing "a longterm project to create a database of research materials on the American South that eventually would cover all chronological periods, subjects, geographical divisions, and types of materials."[11] The proposal's framers envisioned a highly collaborative enterprise involving "faculty, librarians, and appropriate computer personnel [to] develop the policies, strategies, and timetables for the project and raise funds to support it." While the memo itself was too broadly conceived to serve as a blueprint, it did set in motion the follow-up discussions and eventually the research and development projects that became the nucleus of the current site.

The choice of Southern Americana was hardly accidental, building as it did upon the recognized historic strengths of the library's collections and a field of study that enjoyed a surge in popularity during the 1980s and 1990s. It is revealing–although not surprising, given the project's early date–that technological specifics are not detailed in the original proposal. One paragraph lists desired capabilities of the database, such as searchable parameters, and acknowledges the considerable number of technical questions to be resolved.

When the University of Utah's Marriott Library established a digitization center, the goal was to experiment with technology in order to improve access to collections, especially to photographs, rare books, and manuscripts. As early as 1994, the historic photography curator had begun digitizing portions of the photograph collection and placing them on a link-based Web page, but developments at other research institutions soon led the library to think about a more systematic operation. In 1999 the program was formalized under the management of a digital technologies librarian, whose early goals were to explore the potentials of technology for increasing access to rare or unique resources. Kenning Arlitsch, the library's Head of Information Technology, recalls the original proposal that he drafted as "primitive,"[12] but it very clearly lays out the mission of a new digitization center to "provide leadership and means for digitization efforts, primarily within the library and eventually for the campus," and it details the anticipated equipment and personnel requirements.[13]

Following establishment of the department, a Digital Projects Committee was appointed with "lots of ideas, but little idea of how to make them happen." Eventually, Arlitsch, as the Committee Chair, asked a colleague from Special Collections to propose the library's *Harmonia Macrocosmica*, a 1661 atlas of the heavens by Andreas Cellarius. The pre-arranged selection was approved by the committee and the atlas, with its thirty hand-painted color plates, was the first contribution to the formalized digital collections. Subsequent projects–the maps and diaries of the Overland Trails collection, the Tanner Book collection, the Frank Lloyd Wright *Wasmuth Portfolio*, and several small photograph collections–eagerly engaged and even sought technical challenges. In early 2000 the library purchased CONTENTdm, which at that time was still a research project of the University of Washington and used in only a few institutions. In 2001 the library scanned and hosted a small collection of glass plate negatives from the Utah State Historical Society, a collaboration that eventually grew to over 20,000 images and continues to this day. That collaboration also proved the concept of using CONTENTdm to help remote partners create their own digital collections and set the stage for the Mountain West Digital Library and subsequent multi-site projects.

As these brief histories reveal, the broad goal of both projects–to bring digital technologies to bear on analog collections–was essentially the same. The specific goals, however, were quite differently conceived and equally valid. The contrast is perhaps best summed up by Natalia Smith of DocSouth who told us that her approach when she came on board was to envision an interesting project and its features, and then to "make the technology follow."[14] The project team at North Carolina consisted first of bibliographers, with their deep understanding of the collections, and editors, with their sensitivity to the treatment of text, but there was no formal implementation structure. The Marriott Library started with the desire to see what technology could accomplish and a good understanding of the technological requirements. Appropriate preparation there included assembling the equipment and technical expertise before selecting subjects and items within the general classes of materials that could potentially be digitized.

Appropriate Methods

At the heart of any scholarly project is its methodology, a way of proceeding that "gives a project integrity and engenders confidence in its findings, products, or results" (Glassick, Huber, and Maeroff, 28). Methodology can be evaluated through such questions as: "Does the scholar use methods appropriate to the goals? Does the scholar apply effectively the methods selected? Does the scholar modify procedures in response to changing circumstances?" (Glassick, Huber, and Maeroff, 28). Reflections upon methodology were at the start of our own thinking as we began to wonder just what identifies a scholarly project that does not fit the mold of a peer-reviewed publication. What makes these digital library projects qualitatively different from the vast and growing body of facsimiles that are on the Web? Recognizing that we are talking about projects of application that draw from multiple disciplines and fields of inquiry has not made it easier to isolate key features. We have identified two main threads—one related to the norms of humanities research and another to technology—and have thought, as well, about the ways that all methodological choices must be open and defensible. A discussion of methodology also emphasizes the way that initial goals and assumptions influence the evolution and eventual shape of a given digital library.

Documentary Research and Understanding

Unlike the many Web collections that consist of "interesting stuff," we would expect a scholarly digital library project to justify both its topical scope and the inclusion of certain documents rather than others. As one interview subject noted, careful selection was critical in the early days of digitization because the costs were perceived as prohibitive and the long-term costs were unknown. But even as digital conversion has become less expensive and better integrated into library budgets, informed decisions remain a hallmark of scholarly integrity. Digital library developers can approach selection in many ways. For example:

- What is rare or unique?—Both projects have given preference to unique or not widely held materials through the digitization of manuscript items, photographic prints, and similar materials. This focus on rare print and unique primary source materials is often cited as natural role for library digitization programs.

- What reflects scholarly needs?–The Marriott digital library ensures relevance to research and teaching in part by being highly responsive to digitization requests from faculty. For example, a compilation of Julius Caesar's writings, *C. Julii Caesaris quae exstant*, published in 1678 for the use of the French Dauphin, was digitized at the request of a professor in the Classics Department for use with her students. DocSouth is mindful of scholarly trends in ways that follow collection development principles. For example, the program mirrors interests in academia by focusing on social history, personal accounts, and fictional representations, rather than "official" history, although a planned project to digitize the thirty-volume *Colonial and State Records of North Carolina* will constitute a departure from this practice.
- Can an online collection be more than the sum of its parts?–Online collections have the potential to bring together physically dispersed materials in two ways. The first is by aggregating digital collections created and managed by multiple institutions in a single searchable interface. Several Marriott digital library projects implement this approach. Another approach is for one library to digitize materials held elsewhere. When North Carolina set out to pick the first titles for conversion, slave narratives were a logical choice. Not only were the texts rare, but the high circulation rate of copies in the stacks attested to their importance. Early success allowed the project to plan realistically for a comprehensive collection of all known narratives, and some 20% of the texts now included have been borrowed from other libraries to complete the digital holdings. At Utah, interest generated by the Digital Newspaper Project has led to the self-identification of newspaper donors across the state, some of whom have even chosen to donate their print copies to the Marriott Library. Through both these models, digital libraries move beyond the limitations of any one library's physical holdings and open new vistas for researchers.
- What are the library's particular responsibilities?–Although they have taken different paths, both projects we studied made their name largely through a regional focus. DocSouth began with this orientation, which reflected the library's strongest collections; the Marriott digital library came more slowly to regional collections but has realized its greatest successes in terms of grant proposals, interest, and use with projects related to Western Americana. Not all libraries need adopt a regional focus–though public institutions, with their commitment to shed light on local history may find this a

fruitful approach–but all have real or perceived responsibilities that can help shape effective digital library projects.

These are decisions firmly rooted in the traditions of research and of librarianship, especially collection development. They rely upon the acquired wisdom of librarians and their familiarity with the collections in their charge, the ways they are or can be used, and their relationships to the holdings of other libraries and cultural heritage institutions. They must be carefully thought through in order to give the project legitimacy, coherence, and distinctiveness.

Digital libraries must also exhibit understanding of the way research materials are, or can be, used. Most fundamentally, reformatting projects concerned with text must decide whether to present encoded searchable text (the route preferred by DocSouth, based in part on assumptions about the way researchers search and read), indexed images (the approach used in the Marriott digital library's early projects), or a combination (the direction now taken by most Marriott Library projects).[15] Another set of issues involves determining exactly what the object to be digitized comprises. For example, how should features such as covers, indexes, and captions be treated? This is an especially critical question for projects seeking to produce searchable text. Special features and unusual formats pose distinct challenges. The variety of the Marriott digital library's contents has led developers to consider the research needs associated with a wide range of difficult materials. One example is the treatment of Frank Lloyd Wright's *Wasmuth Portfolio*, which contains numerous overlays essential to the understanding of the drawings. A way had to be found to translate the format in order to maintain the full range of its meaning. At the heart of all of these questions is respect for the integrity of the materials, or, as Anne Kenney and Stephen Chapman memorably put it in their advice for digitizing libraries: "Know and love your documents."[16] This respect guides the scholars who consult and interpret books and documents, and it must be embedded in the scholarly digital library's procedures, as well.

One way digital library projects can ensure well-justified decisions regarding the materials is through close collaboration with faculty members. Nearly all DocSouth collections are associated with one or more faculty advisors who take an active role in the selection of specific titles to digitize–usually down to the level of recommending one edition, or sometimes even one copy, over another–and who provide essays and other interpretive information. This structure has been fundamental to the success of the project. Some librarians associated with DocSouth noted that the "First-Person Narratives of the American South" collection, the only project not guided by a faculty member, is the least developed segment. Kenning Arlitsch of the Marriott Library identified strengthening faculty relationships as a priority and expressed regret that faculty members are not more involved in digital library projects there.

Sound Thinking About Technology Standards

The scholarly digital library requires attentive and informed decisions about technology to achieve the goals of enhancing access to collections. Digital library sites must themselves function well and reliably. Further, in today's networked environment, the actual or potential interoperability with external tools and systems is a driving consideration. This is by no means a technical paper and technology changes so rapidly that the available options will never be identical or predictable. Yet, the theme of observing standards where they exist or making a considered decision to modify them emerged repeatedly during conversations with the digital library managers at North Carolina and Utah.

- File formats–Both projects seek to mark up pages consistently and in languages (e.g., XML) that will be recognized by most browsers. DocSouth, with its beginnings as a text project, also made an early and sustained commitment to the standards of the Text Encoding Initiative. Image files need to be created and stored in standard, widely-recognized formats, generally as TIFF files, although they are mostly delivered as JPEG files, which are more convenient for access.[17] Adherence to standards and conventions regarding files ensures that more of the collections will be more available to more users. It facilitates interoperability at the time of file creation or in the future. It also reflects sound preservation practice, as files in common and open or widely supported formats have a better chance of surviving successive generations of migration.
- Metadata–Project managers repeatedly emphasized the importance of sound metadata and of incorporating its generation integrally into digital library development. It is hard at this point to imagine a digital project based in a research library that does not consider metadata standards, but we mention it here because this is one of the features that clearly distinguishes a scholarly digital library from other online collections. It is also an example of the

way that the specific professional expertise of librarians is fundamental to the digital library enterprise.

• Digital Content Management System–The choice of management
system provides a critical opportunity to think about access and
interoperability, although these objectives can be achieved in different ways. Given the early founding date of DocSouth, Carolina
did not have many commercial options available and the library
took the course of building its own database. Because it elected to
use open source software and is compliant with Open Access Initiative (OAI) standards, the system remains flexible and retains the
potential for broad interoperability. The Marriott Library, interested in collaborative statewide projects and benefiting from the
recent availability of several commercial options, chose to purchase CONTENTdm, which promised (and has delivered) interoperability and which also adheres to OAI standards. We see how
factors such as timing and institutional context can influence adoption of two very different solutions, but concerns about interoperability, as well as scalability and cost, will confront any digital
library.

Transparency and Justification

Regardless of methodology, a scholarly project will be open about
the choices made and, when necessary, the rationale for them. Digital libraries needn't be concerned with "reproducibility" in the manner of
pure research in the sciences or social sciences, yet those endeavors provide a model for talking about the relationship of scholarly judgment to
final outcome.

One type of transparency relates to technology choices. The Marriott
Library, for example, publishes a very clear statement about the
digitization process as part of most projects. Readers of the online
C. Julii Caesaris will find the following statement "About the Digitization
of this Book":

> The Caesaris book was photographed with a Leica S1 Pro digital
> scanning camera, mounted on a copy stand and equipped with a
> Hasselblad CFi 50mm f/4 lens. Resulting TIFF files were approxi-
> mately $5,000 \times 5,000$ pixels and 60Mb each, and were resized to
> 600 pixels across and converted to JPEG files for Web delivery.
> Files were loaded into the CONTENTdm digital media manage-
> ment software, assigned Dublin Core metadata, and uploaded to

the library's Web server (http://www.lib.utah.edu/digital/caesaris/about.html).

In providing this information quite visibly in addition to the formal metadata embedded in the file, the digital library permits a researcher to know more about the circumstances of the conversion and consequently about the original artifact and the ways that the conversion process might influence interpretation of content.

Less obvious, but equally important, is transparency about content. DocSouth, by the nature of its subject matter, is especially susceptible to misinterpretation. Selections in the "Library of Southern Literature" include such potentially inflammatory titles as Thomas Dixon's *The Clansman: An Historical Romance of the Ku Klux Klan* and George Fitzhugh's *Cannibals All! or, Slaves Without Masters*. Because the bibliography upon which the Library was built was compiled in an academically sound manner and the bibliography, along with information about its development and an essay on Southern literature, are included as part of the online project, the inclusion of these titles is more readily understood.

Significant Results

We like to think that the proof of significance for both projects resides in the high usage levels and positive feedback that each has continued to generate. We can cite grant awards, media attention, and honors that have accrued to each. We hear in person and electronically from individuals whose research has been enhanced or whose lives have been touched by their encounters with these digitized materials. *Scholarship Assessed* also poses questions–"Does the scholar achieve the goals? Does the scholar's work add consequentially to the field? Does the scholar's work open additional areas to further exploration?" (Glassick, Huber, and Maeroff, 29)–that can help us to think in additional ways about the results achieved.

We know, for example, that the Marriott Library's engagement with CONTENTdm has had a direct influence on the development of that tool and improved its utility to other institutions. Because CONTENTdm's developers were eager to improve their product, Kenning Arlitsch of the Marriott Library reports that his work benefited from a "privileged relationship" that was mutually beneficial. Input from the University of Utah resulted in the expansion of CONTENTdm to manage and search newspapers, for example, and to effectively aggregate content from multiple sites.

The status of both projects as works in progress that are continually being built, refined, and better understood also generates new challenges and avenues of inquiry. One focus of current exploration is the improvement of access, cited by managers of both projects as a priority. The Marriott Library is grappling with ways to improve searching across multiple collections from multiple institutions. "Aggregating collections means losing search granularity," explains Kenning Arlitsch, "because standards are interpreted in slightly different ways at various institutions." DocSouth managers are thinking about ways to better open the collections to many different types of users and queries. A geographic index was just added and others are under consideration. But, as Natalia Smith observes, "We could add hundreds of features and it would just overload our users. How do we come up with ten features that open up the same possibilities as one hundred?" These are practical questions, but they also are indicative of the ways that an application project can itself open up new questions.

Effective Presentation

Scholarship Assessed raises questions about the presentation of results: "Does the scholar use a suitable style and effective organization to present his or her work? Does the scholar use appropriate forums for communicating work to its intended audiences? Does the scholar present his or her message with clarity and integrity?"(Glassick, Huber, and Maeroff, 31). In the case of digital libraries, these questions are particularly relevant, since presentation is integral to the success of the projects, and, indeed, is one of their defining features. Some questions are common to any electronic project: Is the site easy to navigate? Do materials display as intended? Are search tools intuitive or else well explained? Others take on particular urgency because of the nature of the materials and the way they are used: Is each digital object unambiguously identified? Can materials be viewed in a way that is consistent with research needs?

Another aspect of presentation is the visibility of the technology. In DocSouth, the technology is almost invisible. Of course it is obvious that the reader accesses and peruses the material via the Web, but other than that, the search engine, the indexing, and even the metadata work discreetly behind the scenes. In the Marriott digital library, technology is quite prominent. Maps can be zoomed and the seams between two views erased. Photographs can be viewed at several resolutions and there is an ecommerce module for the purchase of high resolution photographs. Newspapers display full page views, transcription views, and even column views. Art images can be thumbnailed and arranged for presentation. All of these capabilities are very much in view and are a key part of the digital library conception.

More prosaically, presentation encompasses the observation of applicable norms and standards. Scholarly projects, regardless of the methodology employed, can be expected to conform to certain conventions. We have examined the attention that sound projects must give to technical standards, and it seems almost tedious to point out that these projects are equally scrupulous, for example, about bibliographic citation. Yet observation of recognized standards does indeed serve as a scholarly identifier, as any style manual will attest.[18] Adherence to standards is both a practical matter, ensuring the functionality and utility of the site, and also an overt sign standing for the rigorous work that has gone into the projects.

Reflective Critique

The final characteristic of scholarly work is "reflective critique," derived through questions such as: "Does the scholar critically evaluate his or her own work? Does the scholar bring an appropriate breadth of evidence to his or her critique? Does the scholar use evaluation to improve the quality of future work?" (Glassick, Huber, and Maeroff, 33). Glassick and his colleagues "found little evidence that this standard figures prominently in the evaluation of scholarship as matters now stand" (Glassick, Huber, and Maeroff, 33-34). The experience of our digital libraries has been quite the opposite. Because these are vital and extensible projects, ongoing reflective critique is critical to prevent stagnation and ensure the programs' optimal utility.

Some mechanisms can be embedded in the project itself. DocSouth's "Contact Us" Web form invites readers to "tell us how you are using *Documenting the American South* and what you think of it." Librarians found the comments so informative that they published a small compilation in 2002 as *Keep Up the Good Work(s): Readers Comment on Documenting the American South.*[19] The Marriott Library's popular newspaper project includes a twelve-question user survey where readers can describe themselves and their research interests and can make suggestions.

It is also possible to critique digital libraries through more formalized channels. Before a new DocSouth Web interface debuted in 2004, the library's information literacy team spent eighteen months conducting usability studies with focus groups of students, faculty members, K-12 educators, and readers from the general public. These studies resulted in an updated "look and feel" and the introduction of features such as a "click-on" map of North Carolina counties (http://docsouth.unc.edu/geographic/ncmap.html). As DocSouth plans to add collections, usability studies are routinely being built into the workflow.

CONCLUSIONS: REALIZING OUR SCHOLARLY RESPONSIBILITY

We have sought in this paper to examine characteristics that our digital libraries share with endeavors more readily classified as "scholarly." As the academic library seeks to redefine itself in an environment of changing expectations and opportunities–an environment in which we will be "more deeply engaged in the fundamental mission of the academic institution," to recall the observation of Wendy Pradt Lougee– then it is by the criteria of academia that we can expect to be assessed. How can our digital libraries help us to align ourselves with and further the mission and values of our institutions? How can they help us effect transformation in our libraries? We conclude with a few observations that might serve to guide us.

First, we need to embrace and actively think about the work that we do as "scholarly" in nature. The academy demands this orientation from its scholars, and enables it by serving as a privileged space for inquiry and experimentation. How can we also think about our libraries in these terms? The values of scholarly endeavor are usually implicit in the research library's role and mission. But to fully realize the library's potential for achievement in the realm of scholarship, we need to make those values explicit within our stated core beliefs and apparent in our most visible public presence, our digital libraries. We need to share among ourselves the conviction that we engage in scholarly pursuits, and we should encourage others to expect this of us, as well.

Next, we need to be acutely aware that we are of our institutions. This means understanding and working within the traditions and organizational context of a given campus. We found that our digital libraries seem to follow patterns established by the library as a whole in its relation to the campus and the faculty. This goes beyond building our digital libraries around institutional academic strengths although, as the North Carolina example shows, this can be very effective. At Carolina, the library has a strong tradition in its collections and services of supporting and extending the University's concentration on the American South, and DocSouth embodies this commitment. At Utah, on the other hand, the library is known for collections in such areas as Mormon history and the settling of the West that, for political reasons, are not curricular strengths but that could nevertheless be expected to fall in the purview of a public institution. Building on this tradition of complementing the curriculum, the Marriott digital library makes available a vast store of primary research information about these topics.

Being of our institutions requires us also to consider and advance special obligations imposed by institutional missions. As flagship public universities, both the University of Utah and the University of North Carolina have a responsibility to serve the people of their states. Our digital libraries provide citizens with direct access to extraordinary collections, often built with public funding, and shed light upon state, local, even personal histories and heritage. At a time when universities are all too frequently charged with being elitist or out of touch, digital libraries can also be a means of building up public trust and respect. This certainly has been the experience at both Utah and North Carolina. "Educating nonspecialists" is one aspect of scholarship identified by Boyer. We need to emphasize this achievement when we talk to our administrators, our legislators, and our users about our digital libraries.

A third useful theme to consider is how we can align ourselves with and learn from other scholarly activities, both within and beyond our institutions. We found it telling that our interviewees, when asked whether digital library development was most like collection development, publishing, research, editing, curatorship, or a technology project, uniformly told us, "All of the above." Just what can we learn from these activities, many of which are unambiguously "scholarly"? DocSouth takes a page from scholarly journals and academic presses by having established at the outset an Editorial Board consisting of librarians, faculty members from a variety of disciplines, technology experts, and representatives from the UNC Press. What other values, procedures, and standards can we import or adapt?[20]

Next, we need to do all of this without abandoning our own values, insights, and contributions. In emphasizing the "scholarly" nature of digital libraries, we don't mean to lose sight of the "library" part, especially our commitments to access, preservation, and service. Credible competitors to our digital libraries exist both within and outside of academia, but as Pat Dominguez of UNC-Chapel Hill told us, "Libraries shouldn't give up because of that. The library is the best organization to take this work on. We have great values. We have great services."[21] We must insist on these values in all that we do. Indeed the positive reaction to our digital libraries, combined with the storehouse of relevant expertise we command, gives libraries a unique opportunity to provide leadership in the adoption and application of information technology for multiple academic purposes.

Finally, if we wish to be thought of alongside more established forms of scholarship, we must be mindful of the deep responsibility this imparts. One quality of a scholar identified by Glassick and his colleagues in *Scholarship Assessed* is integrity. The criteria explored in that book and throughout this paper are all ways of ensuring the integrity of the scholarly process, the reliability of the results. Much of the "scholarly" activity of our digital libraries is, in fact, invisible, especially to readers from the general public, who may stumble upon our projects primarily because they seek information about an ancestor, or because they have broad interest in a topic or because they arrived via a chain of links that cannot even be reconstructed. What they notice, and what serves as a guarantor of accuracy and authority, is the name associated with the content. Our libraries and the institutions of which we are part trade in effect on our institutions' good name. We have a special responsibility not only to our readers, but also to our own institutions, to sustain brand value through the application of core institutional and academic principles.[22] There may be many ways to be a scholarly digital library. We have an obligation to explore them, to integrate those values with our own, and to claim our place for the thoughtful, important, and scholarly work we do.

NOTES

1. Wendy Pradt Lougee. *Diffuse Libraries: Emergent Roles for the Research Library in the Digital Age* (Washington, D.C.: CLIR, 2002), 4.

2. "A Working Definition of Digital Library [1998]," Digital Library Federation, http://www.diglib.org/about/dldefinition.htm.

3. Suzie Allard, "Digital Libraries: A Frontier for LIS Education," *Journal of Education for Library and Information Science*, 43 (Fall 2002): 235.

4. Review performed by Sarah Michalak and Sarah Poteete, University of North Carolina at Chapel Hill Library, February, 2005.

5. Ann Okerson, "Asteroids, Moore's Law, and the Star Alliance," *Journal of Academic Librarianship*, 29 (September 2003): 281.

6. Daniel Greenstein and Suzanne E. Thorin. *The Digital Library: A Biography*, 2nd ed. (Washington, D.C.: Digital Library Federation and Council on Library and Information Resources, 2002), 9-10. Also available online at http://www.clir.org/pubs/reports/pub109/contents.html.

7. Although such arguments certainly can be made. In 1992, Syracuse University launched its national project on Institutional Priorities and Faculty Rewards, which encouraged scholarly societies and associations to develop discipline-based definitions of scholarship. The Association of College and Research Libraries participated in this effort, producing a statement on "Academic Librarianship and the Redefining Scholarship Project." In: Robert M. Diamond and Bronwyn E. Adam, eds., *The Disciplines Speak II: More Statements on Rewarding the Scholarly, Professional, and Creative Work of Faculty* (Washington, D.C.: American Association for Higher Education, 2002), 193-200. Also available online at: http://www.ala.org/ala/acrl/acrlpubs/whitepapers/academiclibrarianship.htm.

8. Ernest L. Boyer, *Scholarship Reconsidered: Priorities of the Professoriate* (Princeton, N.J.: Carnegie Foundation for the Advancement of Teaching, 1990).

9. Charles E. Glassick, Mary Taylor Huber, and Gene I. Maeroff, *Scholarship Assessed: Evaluation of the Professoriate* (San Francisco: Jossey-Bass, 1997).

10. We acknowledge here the difficulty of applying criteria meant for the assessment of individual performance to the activity of an entity such as a library; a certain amount of awkwardness and imperfect fit is inevitable, but we still believe the exercise to be enlightening.

11. Pat Dominguez, Luke Swindler, and David Moltke-Hansen, "Documenting the American South: A Proposal to Create a Multi-Disciplinary, Multi-Media Networked Database at the University of North Carolina at Chapel Hill for Reference, Instruction, and Research on the Region (Preliminary Draft)," 7 March 1992. Manuscripts Department of the Academic Affairs Library of the University of North Carolina at Chapel Hill Records, 1929-1996 #40052, University Archives, Wilson Library, University of North Carolina at Chapel Hill.

12. Kenning Arlitsch, interview by the authors, 21 December 2004. Subsequent remarks attributed to Arlitsch are taken from this interview.

13. "Proposal to Establish a Digitization Center in the Marriott Library," (28 November 1999). Internal document provided by Kenning Arlitsch.

14. Natalia Smith, interview by the authors, 3 January 2005. Subsequent remarks attributed to Smith are taken from this interview.

15. For *Documenting the American South*, which was first conceived in the early 1990s, cost was also a factor. As Natalia Smith told us, the ideal would have been to present both images and encoded text, but the production and storage costs were at perceived the time as prohibitive.

16. Anne R. Kenney and Stephen Chapman, *Digital Imaging for Libraries and Archives* (Ithaca, N.Y.: Department of Preservation and Conservation, Cornell University Library, June 1996) iii.

17. One exception is the decision by the Marriott Library to compress some maps and oversize materials using MrSID software from LizardTech Inc. Once JPEG2000 subsequently became available, it was recognized as a better option for more users. Some files have since been converted from MrSID format to JPEG2000.

18. The very concept of a "style manual" speaks to the weight that the scholarly community places upon standardized presentation. MLA's *Guide to Scholarly Publishing* is explicit on the matter:

Standardization of form keeps you from having to worry about nonsubstantive matters, and as a consequence you can concentrate on your genuinely fresh contributions.

Moreover, observing the codes that have been agreed on within our disciplinary community signals your membership in the community.

Herbert Lindberger, foreword to *MLA Style Manual and Guide to Scholarly Publishing*, 2nd ed. (New York: Modern Language Association of America, 1998), xvi.

19. *Keep up the Good Work(s): Readers Comment on "Documenting the American South."* (Chapel Hill: University of North Carolina at Chapel Hill Library, 2002). Also available online at http://docsouth.unc.edu/about/readers.pdf.

20. We found it interesting to note that both the University of North Carolina at Chapel Hill Library and the Marriott Library at the University of Utah operate small-scale monograph publishing programs. We have no evidence, but wonder if this activity in some ways indicates a predisposition on the part of these libraries to think of themselves as publishers and producers of knowledge, and to be open to integrating the standards and procedures of scholarly publishing.

21. Patricia Dominguez and Luke Swindler, interview by the authors, 30 December 2004.

22. Thanks to Paul Jones of ibiblio for encouraging us to think about the library "brand."

doi:10.1300/J111v46n01_04

New Opportunities for Research Libraries in Digital Information and Knowledge Management: Challenges for the Mid-Sized Research Library

Shirley K. Baker

SUMMARY. Libraries are faced with great opportunities to take responsibility for digital information and knowledge management, both on their campuses and across disciplines. These opportunities come, however, with significant challenges. The challenges are less technical than they are financial and social–identifying funding and penetrating the faculty culture to generate enthusiasm and support for sustainable work. This paper will highlight some experiences, successes, and failures. doi:10.1300/J111v46n01_05 *[Article copies available for a fee from The Haworth Document Delivery Service: 1-800-HAWORTH. E-mail address: <docdelivery@haworthpress.com> Website: <http://www.HaworthPress.com> © 2007 by The Haworth Press, Inc. All rights reserved.]*

Shirley K. Baker is Vice Chancellor for Information Technology and Dean of University Libraries, Washington University, St. Louis Libraries, Campus Box 1061, One Brookings Drive, St. Louis, MO 63130 (E-mail: baker@wustl.edu).

[Haworth co-indexing entry note]: "New Opportunities for Research Libraries in Digital Information and Knowledge Management: Challenges for the Mid-Sized Research Library." Baker, Shirley K. Co-published simultaneously in *Journal of Library Administration* (The Haworth Information Press, an imprint of The Haworth Press, Inc.) Vol. 46, No. 1, 2007, pp. 65-74; and: *Digital Information and Knowledge Management: New Opportunities for Research Libraries* (ed: Sul H. Lee) The Haworth Information Press, an imprint of The Haworth Press, Inc., 2007, pp. 65-74. Single or multiple copies of this article are available for a fee from The Haworth Document Delivery Service [1-800-HAWORTH, 9:00 a.m. - 5:00 p.m. (EST). E-mail address: docdelivery@haworthpress.com].

KEYWORDS. Innovation in higher education, innovation in libraries, digital projects, academic libraries

INTRODUCTION

Not all digital information or knowledge management efforts are successful. Other writers will certainly address lofty visions and stunningly successful projects. But, sharing information about failures or partial successes can also be useful for the library community. Thus, this paper addresses efforts which did not always succeed as initially imagined.

On one hand, we are living in times of breathtakingly rapid change. How could we have imagined that so much scholarly information would become available electronically so quickly? Or that so many "timeless" practices could become superfluous? And yet, for the past decade, I've felt that, no matter how hard we try, change comes at its own plodding pace. When I read the latest edition of *EDUCAUSE Review*, I marvel at the developments but I often think, "Be prepared to wait a while to see that here." What is going on?

RESEARCH ON THE SPEED OF INNOVATION

So I revisited what research tells us about change and innovation in higher education. Often, as I've mulled over opportunities and obstacles, I've thought of Malcolm Getz's research. Most people now know Malcolm Getz as a Vanderbilt economist. But some of us knew him as the director of Vanderbilt's Libraries in the late 1980s. His research has focused on the speed of adoption of technology. He studies it in industry and in higher education. In 1997 Getz and two colleagues published the results of some of their work as *Adoption of Innovations in Higher Education.*[1]

Getz et al. measured the adoption rate for thirty innovations in six categories in 238 colleges and universities in the United States. They compared higher education innovation adoption rates with those in industry. There is evidence from industry that more highly educated managers adopt innovations faster than those less highly educated. Thus the authors posited that colleges and universities, rife with highly educated managers, would innovate faster than industry. And, they test an assertion that higher education is innovative because of the diversity of institutions involved. If this is so, then diverse higher education should innovate more quickly than industries that are more uniform.[2]

THE STUDY

The six categories that Getz et al. investigated were *curriculum, classroom services, student life, libraries, computing and telecommunications*, and *finances*. For each of the six, the investigators chose five innovations. The innovations they chose had to have been implemented recently enough to be remembered, sufficiently well accepted to be judged a success, and they had to be applicable to most higher education institutions. The five innovations tracked for libraries were *bibliographic utility participation, database searching, optical disk use, automated circulation*, and *an automated public catalog*. These innovations are far enough in the past to be ancient history to librarians. But they are recent enough that we can identify when we adopted each.

OVERALL RESULTS

For the six higher education area studied, innovations required an average of 26 years. This time period is measured from the first adopters until at least half of the institutions have made the innovation. By contrast, in similar studies of innovation in industry, the time for the same level of penetration was eight years–more than three times as fast as colleges and universities. The for-profit sectors examined were manufacturing, agriculture, and health care.

Quite obviously, the level of education of managers is not driving a fast pace for innovation in higher education. Nor does our diversity seem to give us an edge.

RESULTS BY AREA

Within higher education, the authors compared the rates of diffusion of innovation in their six categories. They measured the number of years from when the first ten percent had adopted until the median of the group had adopted the innovation.

Area	Years
Financial services	15.0
Classroom services	12.6
Curriculum	12.4
Student Life	9.7
Libraries	4.8
Computing	4.8

Why do computing and libraries adopt innovations so quickly, at a rate similar to industry? I am reminded of an anecdote about libraries and innovation often told by the James Joyce scholar Hugh Kenner. He talked of having made his first photocopy in a library. Kenner first put fingers to a computer to search the online catalog in a library. And, he avers that James Joyce saw his first electric light bulb in the Dublin public library.[3]

The study authors posited that possibly library and computing innovations might be more visible than those in other areas. However, this was not supported. The computing and library innovations were no more visible than other innovations that happened much more slowly. Similarly, pressure from vendors did not appear to be a cause. Although vendors certainly market to computers and libraries, innovations were also marketed to others, but not adopted. The study authors do posit that the pace of change in technology does help drive library and computing innovations. And, they posit that because libraries and computing are more hierarchically structured than other areas of universities, decisions can be made more quickly, without achieving consensus. And, the authors posit that library and computing leaders may be more professionally active and therefore more aware of developments in other institutions. We librarians do tend to know what the *Jones'* are doing and we try to *keep up with them*. But, while we are competitive, we are also very willing to share information and expertise, a surprise to vendors new to the library business.

OBSERVATIONS

So, if Getz' findings are true–and I think they are–why the sense of plodding? Change *is* happening fast in libraries. We have made many innovations since those studies by Getz et al.–e-reserves, e-collections, self-service ILL, patron-initiated borrowing, digital image databases, federated search engines, institutional repositories . . . examining the last of these, institutional repositories, may shed some light on the problem. These repositories require collaboration–not collaborating with other librarians (we are good at that) but collaborating with faculty.

We librarians take our professional obligations seriously and act accordingly. But we often act on those obligations years before they become compelling to our faculty colleagues. Many of our library innovations are met with befuddlement or with grudging tolerance or even with outright complaint by significant portions of our academic communities.

COLLABORATION WITH FACULTY

What happens when interaction with the faculty culture must occur? Hurry up and wait. Even the adoption of e-reserves and course management systems took considerable time to penetrate faculty culture in many institutions. There are always some faculty early adopters, but student pressure becomes an impetus for many who might have preferred to continue previous, comfortable, and known practices.

We librarians have been talking about the need to capture and preserve digital assets since at least 1995. In that year Michigan's Margaret Hedstrom wrote an article on the topic with more than 20 pertinent references.[4] And, MIT rolled out its DSpace in 2002, after several years of development work with grant funding and industry support. Other initiatives appeared about the same time. And we now have multiple choices for repository platforms, should we (perhaps wisely) choose not to write our own.

But faculty response to institutional repository capacity has been less than overwhelming.[5] A reporter for the *Chronicle of Higher Education* wrote about this in a late 2004 article. According to that article, MIT has now mounted a local public relations campaign to more attract contributions from their faculty. Some libraries–Georgia Tech, Cal Tech, and the University of Toronto–have taken responsibility for themselves collecting and depositing faculty creations. And the same funding organizations that supported the development of depository systems are now funding anthropological studies of faculty behavior (at the University of Rochester) to match need with capability.[6] This is an important occurrence. We need to understand where, in our faculty members' very busy lives, we can insert low-barrier opportunities for them to use our services for their benefit.

ANALYZING SUCCESSFUL AND LESS-THAN-SUCCESSFUL PROJECTS

At Washington University, we have been working on some digitization projects but have suspected that they compared unfavorably with the range and scope of other universities' work. To test our suspicions, we have surveyed large digitization projects–the ones that contribute to our feeling of inadequacy. We have also analyzed our own projects in that light. Our findings for the two quests reinforce each other and allow us to tease out some of the key issues.

Many libraries' large digital projects have three common characteristics: the libraries own the collections chosen, the collections are of interest to a broad range of users, and there are no complications with intellectual property rights. And, with few exceptions, the projects were library driven and library controlled. Good examples of such highly successful projects are the University of Utah's *Utah Digital Newspapers* project (www.lib.utah.edu/digital/unews/index.html) and the University of North Carolina's *Documenting the American South* (www. docsouth.unc.edu).

WASHINGTON UNIVERSITY EFFORTS

Our digital efforts have primarily been undertaken in concert with faculty. We felt that we would best serve the University with that approach. So, in the late 1990s, we created a faculty/librarian group to research digital needs on campus. We had fourteen faculty from across the university involved in the process. We found great interest but also identified equipment and support needs in the schools and departments. We did fund three experimental projects–fashion drawings from a design professor, an anthropologist's images of a Japanese fishing village, and some early work with maps and GIS. We created a Website to guide librarians and faculty in planning and implementing future digital projects. But our efforts stalled. The experiments came to little as we struggled with very modest academic support and the then-limited capacity of our software.

But some of our humanists began to get a whiff of what their colleagues at other institutions were doing. (Some were feeling the long shadow of the University of Virginia's *Valley of the Shadow* (www. valley.vcdh.virginia.edu). So they came to us to see if we could work together. One group had identified an amazing resource in the community that–if it could be made digitally available–would be of international importance. These were the St. Louis Circuit Court records–a complete run from 1789 to the present day. They were untouched, still in their legal tri-fold form. There had been no floods, no fires, so everything was there. And St. Louis played a major role in the westward expansion of the country–think Lewis and Clark and the Oregon and Santa Fe Trail and 100 years of up-the-Mississippi migration.

Working with the Court, the Missouri State Archives, and our faculty, we investigated the records for good pilot projects. The most obvious choice was the original court documents from the Dred Scott freedom suits. (Dred and Harriet Scott won their freedom in the St. Louis courts but the verdict was reversed by the U.S. Supreme Court, adding to the turmoil that led to the Civil War.) We librarians played a major role in the pilot project, digitizing and transcribing the original and mounting it on a Web site (www.library.wustl.edu/vlib/dredscott). That project got international attention and–years later–the documents are among the top three things searched on our Website.

Then we began planning for the next stages of the project. First we would focus on the many lawsuits involving Meriwether Lewis and William Clark. (After the expedition returned to St. Louis, everyone sued everyone else.) Then we would digitize the many other freedom suits in the records (www.stlcourtrecords.wustl.edu). The Scotts chose to bring their case to court in St. Louis because the city had a good record of awarding in favor of those seeking their freedom. But then, we and our faculty partners began to have differences of opinion in how to scale the projects.

We had many discussions about software platforms and metadata descriptive standards, often at cross purposes. The Libraries favored using an existing software platform, one that would be maintained over time. However, the faculty leaned toward using simpler methods cobbled together by their more-computer-literate graduate students. And, our recommendations for descriptive data were considered far too time consuming apply. Our concerns about scalability and preservation over time were considered stumbling blocks to fast progress on an interesting project. So the faculty forged on independently.

We, however, thought we could have tried harder on this faculty project. We continued to look for ways to be helpful. Twice, when the project ran into obstacles, we revisited the issue and were able to assist modestly. But we weren't getting very far. We offered assistance with a second humanities faculty project–one involving oral histories, images, sounds, newspaper coverage, and archival materials (a snake pit of intellectual property issues). Our offers were once again rejected as being too obstructive and not sufficiently customized. But not having developed a robust capacity for straightforward digital projects, we were ill-equipped to engage in the desired research and development activities–especially when they involved undergraduates writing new search engines.

The ante was raised in the library-faculty collaboration sweepstakes when the Libraries acquired a film archive of vital interest to influential faculty. This was the archive of documentary filmmaker, Henry Hampton, Jr., and his company, Blackside, Inc. Hampton made the groundbreaking documentary series *Eyes on the Prize*, which documents the heroic efforts of ordinary people in fighting for their rights. Blackside also made several more series–on the great depression, on Malcolm X, on African American artists and scientists. Hampton's archive is a goldmine of often unique materials for research on the history of the 20th century.

Dreams of research and teaching incorporating digitized sections of the archive drove faculty back to the discussion. (We have those same dreams. But our dreams are colored by the three semi-trailer-loads of archival materials in dire need of organization and preservation.) Thus, a new collaborative effort–the Digital Archives Task Force–was born. We began with some education. We looked at successful faculty efforts on our campus–in science and the social sciences.

We did some readings and attended some conferences. The wisdom of Clifford Lynch began to be acknowledged. We spent some time discussing what Lynch refers to as "the dirty little secret"[7] that many highly-touted digital projects may not be scalable or sustainable. All parties affirmed the importance of (at least) minimal metadata. And, we librarians worked at unlearning librarian-speak so that we could be intelligible to faculty. We re-did the digital projects Web site to make it accessible to non-librarians (www.digital.wustl.edu).

We accepted the Arts & Sciences model of *tiger teams* for technical projects. The *tiger team* of librarians, slide curators, and art history faculty succeeded in creating a digital archive of images using the new Visual Resources Association (VRA) standard for description and off-the-shelf software (www.library.wustl.edu/subjects/art/luna.html). We are supporting an increasing number of large image-based classes and, with the recent addition of ARTstor to our licensed resources, are in good shape for delivery images for teaching.

CONCLUSIONS FOR WASHINGTON UNIVERSITY

So our projects don't always fail. Where do we succeed? Well, we succeed when we work with other libraries–public or academic. We have underway a successful project for digitizing St. Louis Sanborn maps, involving a public library and two special libraries. We succeed when working with faculty in disciplines where collaboration is the norm. The model of the humanities scholar working in isolation plays against successful collaboration. They just don't have the skills or practice. And, as MIT's Jay Lucker used to say, "Collaboration is an unnatural act."[8]

We work successfully with faculty who themselves are technically sophisticated–those who are at least as knowledgeable as their graduate students. Graphic artists are some of the technologically sophisticated on our campus. (We do try to not be crushed by their opinions of librarians' design sense.) Computer scientists are difficult for us librarians to work with; our interest in on-going service conflicts with their interest in the new and untried. With some projects, we sometimes succeed because we are providing the funding and can "buy" agreement on issues of sustainability and scalability.

On reflection, we see that we have had some failures and some successes. We have determined that we can no longer *ad hoc* projects by exploiting the interest and skills of existing staff who already have other full-time assignments. Thus we are working on building a robust infrastructure, through a combination of reallocating and arguing for new University funds. We are repurposing staff–choosing not to do things previously considered sacred. We will be buying some shelf-ready books, so that we can concentrate our cataloging expertise on metadata and film cataloging. We are eliminating paper journals when we get electronic–saving serials check-in and binding staff. We are consolidating services points and making many other changes common in other libraries. Our case for new University support will be based on a commitment to continue to work with faculty, sharing positions and collaborating on projects. We still think that collaboration will bring the best results for our institution, if we keep at it until we get it right.

NOTES

1. *Adoption of Innovations in Higher Education*, Malcolm Getz, John J. Siegfried, and Kathryn H. Anderson, *Quarterly Review of Economics and Finance*, Vol. 37, No. 3, Fall 1997, pp. 605-631.

2. In the survey of earlier work, the authors quote a 1979 Getz study of innovation in fire departments. His findings are quite alarming. Fire departments studied took thirty to forty years for complete diffusion of innovations which did not involve the handling of water. They took seventy to 100 years for those that did.

3. Heard in person sometime in the early 1980s and read in a Research Libraries Group newsletter at a later date.

4. Hedstrom, Margaret. *Digital Preservation: a Time Bomb for Digital Libraries.* (http://www.uky.edu/~kiernan/DL/hedstrom.html).

5. *Papers Wanted: Online Archives run by Universities Struggle to attract Material*, Andrea Foster, *Chronicle of Higher Education*, January 25, 2004.

6. *Understanding Faculty to Improve Recruitment for Institutional Repositories*, Nancy Fried Foster, Susan Gibbons, *D-LIB Magazine*, January 2005, Vol. 11, No. 1.

7. Oral comment in Clifford Lynch's closing remarks at *Institutional Repositories: A Workshop on Creating an Infrastructure for Faculty-Library Partnerships*, Co-sponsored by ARL, SPARC, and CNI, October 18, 2002, Washington, DC.

8. Heard often in the seven years (1982-89) that I worked with Lucker at MIT.

doi:10.1300/J111v46n01_05

Is There a Digital Purgatory?

Charles T. Cullen

SUMMARY. The many opportunities libraries face with the advances of digital technology lead to an understandable assumption that providing a wide range of resources in electronic formats is a natural next step. Only now are some beginning to give serious attention to problems we need to solve before we move forward too quickly and create more problems for those managing information. This paper addresses these issues from the perspective of a research collection focused on scholarly use. doi:10.1300/J111v46n01_06 *[Article copies available for a fee from The Haworth Document Delivery Service: 1-800-HAWORTH. E-mail address: <docdelivery@haworthpress.com> Website: <http://www.HaworthPress.com> © 2007 by The Haworth Press, Inc. All rights reserved.]*

KEYWORDS. Historical editing, automation in the Newberry Library, knowledge management, digital information, digitizing special collections

Charles T. Cullen is President and Librarian, The Newberry Library, 60 West Walton Street, Chicago, IL 60610-7324 (E-mail: ctcullen@newberry.org).

The author would like to express thanks to James Akerman, Tobias Higbie, Douglas Knox, John Long, and Melissa McAfee, colleagues at the Newberry Library who shared their thoughts in conversations and e-mails, some of which he used in this paper.

This paper was presented at a conference on Digital Information and Knowledge Management: New Opportunities for Research Libraries, March 3, 2005, Oklahoma City, OK.

[Haworth co-indexing entry note]: "Is There a Digital Purgatory?" Cullen, Charles T. Co-published simultaneously in *Journal of Library Administration* (The Haworth Information Press, an imprint of The Haworth Press, Inc.) Vol. 46, No. 1, 2007, pp. 75-88; and: *Digital Information and Knowledge Management: New Opportunities for Research Libraries* (ed: Sul H. Lee) The Haworth Information Press, an imprint of The Haworth Press, Inc., 2007, pp. 75-88. Single or multiple copies of this article are available for a fee from The Haworth Document Delivery Service [1-800-HAWORTH, 9:00 a.m. - 5:00 p.m. (EST). E-mail address: docdelivery@haworthpress.com].

Each of us comes to the issue of digital information and knowledge management from our own particular perspective. Mine has been shaped by my experiences as a practicing scholar at the time the computer emerged as an important tool for processing words and information.

My scholarly focus has been on documentary editing which I did full-time for sixteen years before I began my current position at the Newberry Library nineteen years ago. As a documentary editor and scholar, my interest in the computer was great, and I used it in my daily work of research and editing as much as possible. When I shifted from practicing scholarship to managing it my focus changed considerably. It broadened out of necessity as I learned the importance of looking at the big picture while helping to advance a research library and educational center. My first direct experience with the computer had begun in the mid-'70s when I was intrigued by the possibilities and promises offered by the machine to shorten the amount of tedious work I had to do as an editor in compiling an index to a rather thick volume of documents every eighteen months or so. This was long before anyone thought of digitizing the text and having the computer select key words to compile into an index that would be printed at the back of the finished volume. I used the mainframe computer simply as a fast way to sort my index cards and to produce a file that could be edited quickly and submitted to the publisher as a printout for composition and printing. The technology moved more rapidly than publishers as the personal computer was introduced early in the '80s, and within five years publishers began to recognize the time saving advantages of accepting manuscripts in electronic form. In these early years people were less interested in considering the file that produced type a database–that is, a valuable piece of digital data that contained a great deal of information. This recognition was not long in coming, however, as scholars acquired their own PCs and the search for information to turn into knowledge led to the increasing recognition that more could be done with the digital information than the printed matter. But this required a change in our culture, one that took only another decade or so to advance to a point where PCs are as common as toasters and Google is a common everyday tool.

When I arrived at the Newberry Library in 1986 the Library owned no personal computers, and the few owned by individual adventurous staff were restricted in their use. All letters leaving the Library were to be prepared by the typing pool using Displaywriter word processors. This was a primitive, and unwise, attempt at managing digital information. It was thought that a central repository of all correspondence kept on disc in one office would be the best way to preserve the files for future retrieval and access. As a documentary editor, I had gone on record opposing the use of word processors for any use at all. I favored the adoption of personal computers because of their greater range of functions that would allow users to expand applications into whatever realm of this new world beckoned them. I eliminated the typing pool and encouraged the adoption of personal computers throughout the Library. Today, like everyone else, we are fully computerized with a state-of-the-art intranet, and we are facing consideration of all of the interesting and sometimes perplexing questions that bring us to such conferences as this. And we are looking for answers along with everyone else.

One more bit of personal information will finish explaining my biases. While I was editor of the Papers of Thomas Jefferson, I came upon an interesting document that needed more than the usual amount of research to explain it. It was a rather short list of letters Jefferson received while he was Secretary of State, and the list was obviously selective because it included letters received between 1789 and 1793, too long a period for such a short list if it were anything other than selective. My first thought was that the ink might reveal whether the list had been made over a long period of time (with the varying intensity suggesting the use of different inks at different sittings) but I had only a photocopy from which to work and the contrast was insufficient to enable me to determine differences. I therefore went to the Library of Congress to look at the original, and I saw rather quickly that the list was made at what appeared to be three different sittings. This was significant because it suggested to me that Jefferson had made this list for specific reasons rather than merely as a catalogue, but observing that it was written in three sittings did not tell me when he might have made the list. For some reason I still do not understand (it must have been divine intervention) I lifted the paper toward the ceiling and looked at it with the overhead light behind. It revealed a watermark with the date 1804. This was very exciting to me because it was hard evidence that the list was made more than a decade after Jefferson left the office of secretary of state. Further research indicated rather quickly that Jefferson used paper made with this watermark only into 1805, which narrowed the timeframe he might have made the list. I then remembered that John Marshall published his biography of George Washington in 1804 and I knew also that Jefferson was troubled by Marshall's interpretation of his role in Washington's administration. Reading the contents of these letters listed in the Jefferson documents then made me aware that each in some way defended Jefferson's actions as secretary of state and, in his mind at least, served as justification for his differing interpretation of that period.

My experience with that document has shaped my thinking about documentary editing in a rather profound way. I believe it has also influenced my thinking about digital information and knowledge management in ways that have been beneficial to me in my work.

In an address on "The Book in Globalization Times" Tomas Eloy Martinez, an Argentine novelist and director of Latin American Studies at Rutgers University, described the "Kingdom of Virtuality" as a new form of library. He said it had "given us back, in a certain way, the communal way of reading, of communicating and interacting through signs." He calls it also a "library of Babel" and a "new form of agora, that purgatory or heaven of virtuality." After exploring the evolution of the book and what some regard as the decline of knowledge, Martinez ended his speech with the following sentence: "We, here and now, still imagine Paradise as some kind of library" (IFLA conference proceedings, Buenos Aires, August 2004).

The title of my remarks today was not influenced by Martinez and his musings on the book, but I find his references to the two circles of Purgatory and Paradise to be interesting because I think he and I might agree on where we might find ourselves as we deal with the issues that have brought us together today. I doubt many of us would disagree with his concept of Paradise as some form of library, but I doubt also that many of us would think of even the virtual library Google has planned as being in the circle of Paradise. I see it as falling into the circle of Purgatory because I believe there are serious negatives as well as desirable positives that keep it from falling either into Hell or into Paradise. I would like to explore with you some aspects of the current world the new technology has created for us as we think about opportunities we face in managing digital information and what we hope is the resulting knowledge. I am confident that many of us who will be making presentations to you during the course of these two days will touch on some of the same points. I believe also that there is more likelihood we will find more areas of agreement than difference, and this in itself is evidence, to my way of thinking, that we are beginning to deal with digital information and knowledge management in a constructive manner.

I think the essence of what I have to say can be encapsulated into just a few points, and perhaps into one.

- Too much digital information, too little knowledge management
- Computer technology has made it too easy to produce data and information (and perhaps even knowledge)
- Librarians (some at least) are here to stay

Surely there is overlap among and between these three points, but I think we would all agree that there is a rapidly increasing amount of information available in digital form that necessitates something like Google to help us navigate through it. We can all relate horror stories of misuse by students and others as they try to find information about various topics. Teachers have had to devise methods to ensure students do not cut and paste for their term papers. Each of them is also challenged to make some headway in convincing students to question the reliability and value of information they encounter as they seek material to help them make a point. I confess that I too am an avid user of Google and the Internet to obtain information about almost anything that comes to mind. Like others, I suspect, I immediately turn to Google when I want to know a fact about anything, from information that will help me solve a crossword puzzle clue to biographical information on a scholar or donor prospect, to flight information, weather information, interesting things to do while I am in Australia for a meeting later this month, humanities discussion groups–the list is almost endless, and the rest of my time in the computer is most often spent with e-mail, maintaining important business and personal communications in a manner I was unable to do ten years ago.

As an early Google user, I shared the excitement generated by its recent announcement that it has formed agreements with several outstanding research libraries to digitize their entire collections. I am sure all of us had our own reaction informed by the work we do in libraries and with computers. We all read editorials in the press about this brave new world. Daniel Akst writing in the *Wall Street Journal* praised Google's commitment to "spend millions to break down library walls." Stanford's librarian declared, "Within two decades most of the world's knowledge will be digitized and available, one hopes, for free reading on the Internet, just as there is free reading in libraries today." Predicting the end of printed books on paper as he exuded about this coming Paradise, Akst ended with a thought about the meaning of this for librarians. He predicted that their services would be needed even more as the physical library goes away, remaining only as buildings with lions out front, soon to be relegated to the status of museum, according to Larry Lannom, the Director of Information Management Technology at the Corporation for National Research Initiatives. But Lannom allowed that "the skills librarians bring in organizing and making accessible the ongoing cultural records of humanity will probably change, but they will endure." The *Wall Street Journal* editorialist opined that in the new information economy librarians may command such high pay that people will go into the profession for the money! (David Akst, "Strangers in the Stacks," *Wall Street Journal*, Dec. 17, 2004, W15). Paradise, indeed.

Those of us who work in the field of information management recognized immediately that Google's plan was full of complicating factors, not the least of which was our intricate copyright law. I am told that just last week the brakes were being put on the enthusiasm that came in the wake of Google's announcement. Not surprisingly, there is talk of charges to access these books that will be in the project in digital form. Surely some library staff member at one of the participating institutions thought of this before the announcement was made public, or perhaps in the press's enthusiasm in making the announcement the problems of managing such a digital repository were glossed over, leaving those of us who meet to talk about such things with plenty to question about the project at large.

We come back, then, to recognizing what might appear to be Paradise as more likely a form, and a desirable form, of Purgatory, something between the best and the worst and something that is more realistic for us to contemplate in our brave new world of rapidly advancing technology. In sum, we cannot have any amount of digital information that is meant to contribute to our knowledge without having from the beginning some amount of management that will help users go from data to information to knowledge to wisdom.

A discussion on the Humanist Discussion Group last month comes to mind. Someone from King's College in London began a rather philosophical discussion on the question what is knowledge. The question was generated by the quotation, "data is not information, which is not knowledge, which is not wisdom." The writer asked practicing humanities computing scholars whether that discipline had changed people's perceptions with regard to the distinction among these terms. One bright respondent from Missouri suggested that computing has in fact changed people's perceptions about many things, and he dreaded that computer-mediated communication had reduced "all discourse to information." He cited a physicist who had written:

> Some intellectual prophets have declared the end of the age of knowledge and the beginning of the age of information. Information tends to drive out knowledge. Information is just signs and numbers, while knowledge has semantic value. What we want is knowledge, but what we often get is information. It is a sign of the

times that many people cannot tell the difference between information and knowledge, not to mention wisdom, which even knowledge tends to drive out. (1988, 49)

If our traditions cannot keep knowledge and wisdom alive, these distinctions will disappear as all is reduced to information. The cyborg's spiritual quest would become an endless search for the information that saves–a quest doomed to failure, an endless and eternally restless manipulation of signs and numbers that, like the search for the philosopher's stone, can never produce the gold or the semantic value that we seek. When the ambitious dream described by Richard Lanham in *The Electronic Word* is realized, and the whole record of human culture is digitized and available on computer databases connected to each other by a global web, our spiritual crisis will remain and even intensify, for we will be forced to confront the fact that no electronic alchemy can turn information into knowledge, or into the wisdom that will teach us how to live. [Pagels, Heinz. 1988. *The Dreams of Reason.* New York: Simon and Schuster. Cited in O'Leary and Brasher, *The Unknown God of the Internet: Religious Communication from the Ancient Agora to the Virtual Forum,* in Ess (ed.) Philosophical Perspectives on Computer-Mediated Communication (Albany, NY: SUNY Press), 262]

I am going to risk delving a bit more deeply into this philosophical concept by sharing with you a bit more from this recent exchange. The writer decided to reflect on the meanings of the four terms and the distinctions between them, and he offered for initial thoughts the following:

- data–bits (1/0s) as recognized and manipulated by computational devices.
- information–data organized into both basic and complex units (e.g., the boiling point of water is 100 degrees centigrade).
- knowledge–(human) awareness of units of information and their interconnections. . . .
- wisdom–. . . informed knowledge of how to live well/appropriately as a human being in a human/natural community.

I will spare you the very deep philosophical discussion that followed from a reader in New Castle, Scotland, but the T. S. Eliot lines from his "Choruses from the Rock" are worth reading as they might offer us

some food for thought as we consider digital information and knowledge management:

> The Eagle soars in the summit of Heaven,
> The Hunter with his dogs pursues his circuit.
> O perpetual revolution of configured stars,
> O perpetual recurrence of determined seasons,
> O world of spring and autumn, birth and dying!
> The endless cycle of idea and action,
> Endless invention, endless experiment,
> Brings knowledge of motion, but not of stillness;
> Knowledge of speech, but not of silence;
> Knowledge of words, and ignorance of the Word.
> All our knowledge brings us nearer to our ignorance,
> All our ignorance brings us nearer to death,
> But nearness to death no nearer to God.
> Where is the Life we have lost in living?
> Where is the wisdom we have lost in knowledge?
> Where is the knowledge we have lost in information?
> The cycles of Heaven in twenty centuries
> Bring us farther from God and nearer to the Dust.

–T. S. Eliot, "Choruses from the Rock"

I leave this part of my reflections with the observations that I found
all of this in digital form on the Internet, and I will share with you the
last thought of the correspondent who shared the Eliot poem: "T. S. Eliot
as a forerunner of humanities computing? Oh that happy thought. . . ."
(See http://lists.village.virginia.edu/lists_archive/Humanist/v18/0537.html,
and 0545.html, 0551.html, and 0555.html for this discussion.)
        Those of us who work with scholars in research libraries have re-
flected recently on the ease of creating digital products. It used to be that
one would conceive of a research project, conduct the research, and produce a manuscript that needed publishing for wide-spread dissemination. The first task then was to find a publisher. Now many do it on their
own. The implications of this may be profound. The several roles of
publisher can be bypassed easily, not the least of which is that function
of vetting manuscripts that leads those who find printed books from
publishers in libraries to assume that someone in a respected position of
authority and responsibility has deemed the material within the two
covers worthy of dissemination. The library acquisitions staff have de-

cided the material is worth paying for. It is assumed also that permissions to use others' materials like photographs or lengthy quotations have been obtained, and no copyright infringements exist in using the material. If we as institutions are to become our own publishers, we have to confront the problems we create for ourselves in assuming these duties. I can think of few excuses for research libraries to permit the movement down the road toward digital publication of any product without finding ways to maintain standards in the first instance, and to consider the multitude of issues that should be confronted when a digital product is produced. Because of the problems many of us are recognizing in producing digital products, it is rapidly becoming evident that collaboration is almost always desirable, and perhaps it is even required. It is easy for an individual to put the result of some research on the Web, but for it to be something that would be as valuable as a book that has resulted from serious and high quality research, it simply needs more than transference into bits and bytes. What can be said about this kind of project applies to group projects that we are seeing with more frequency (and understandably so).

Digital objects should be created in a way that ensures they will persist over time despite changing technologies. Even photographic images that are meant to be viewed in low resolution on the Internet should have a very high resolution master file that can be kept up to date for the creation of future slave files as technology evolves. Interoperability is important. Metadata is necessary. The description should be based on best practices and national standards, and it should facilitate discovery of the collection while helping the user understand the nature of the collection. Collection level records should be input into national union catalogues, the library's local catalogue, and listed in national directories of digital collections. Selection at the institutional level is another important issue. A decision to create a digital collection should follow from the library's collection development policy. To avoid duplication of effort, data bases of high quality digital images should be consulted and used as models. Sustainability can be a costly but essential part of digital projects. This includes the persistence of the digital object (the image) and the metadata; access (search systems and applications) should remain usable over time, updating software is necessary. And in order to facilitate the questions of authenticity, security of our servers has to be ensured. The attitude too often seen of "just digitize it and get it out" should be a caution to us all.

Permit me one quick example before I go on to other thoughts that might illustrate in part what I think all of us would recognize as a continuing problem. As I mentioned earlier, my background includes a period of time as the editor of Jefferson's Papers. Since 1943 millions have been spent on ensuring that accurate texts are available to anyone who wishes to consult Jefferson's writings. This project began the modern practice of scholarly documentary editing as it was founded by the first editor of Jefferson's Papers, Julian Boyd. He observed that the many editions of great American's papers were seriously flawed. First, they were highly selective and in almost every instance included only the letters and other writings of the subject himself (and, yes, most of these people were men). Boyd declared that both sides of the correspondence needed to be published in order for Jefferson's letters to be understood in context (and after many years of editing his letters, Boyd concluded that Jefferson often wrote what he thought people wanted to hear, shedding very important light on Jefferson's letters and helping those who want to use them understand them far better than if they were taken in isolation). What Boyd was doing in the '40s as he compiled a huge collection at Princeton of copies of every known Jefferson letter from anywhere in the world would be thought of today immediately as a source for a wonderful digital file. The current world of the Internet and secure servers seem tailor made for such a project. But consider the relative ease Professor Boyd had in working in his study preparing Jefferson letters for publication in book form compared to what a single scholar with one or two assistants would have to do today to prepare the same material for publication in digital form. Before we think of that further, let me hasten to say that early in our modern digital world some enterprising soul decided to scan the volumes, twelve in all, of a 1902 edition of Jefferson's writings, incomplete and selective as they were. Moreover, the transcription of the selections is highly flawed. But this is the file that is most widely available today on the Internet, and the file is available without a single caution to users. We all can recognize the problems presented when students or curious readers turn to this database and read what they think Jefferson wrote in this flawed state.

With my background both as the editor of Jefferson's Papers and the President and Librarian of the Newberry Library, it is not surprising that I might argue strongly that collaboration is practically a necessary component of managing digital information and knowledge.

The preparation of a collection of digital information that may be used for the acquisition of knowledge is a complex undertaking. But I think of several examples where knowledge management is actually enhanced by the wonderful world of digitization. One brief example may suffice, and it comes once again from Jefferson's writings. In 1784, Jefferson published the only book he ever wrote, *Notes on Virginia*, while he was in Paris. For the rest of his life he edited this manuscript, crossing out and inserting revised or new thoughts, tipping into his bound manuscript bits of paper on which he had written additions and emendations. The surviving manuscript is owned by the Massachusetts Historical Society, and it has been published in several editions since Jefferson's death in 1826. The compilation of the manuscript as it existed when he died is a documentary editor's nightmare, because trying to insert emendations and note corrections or the fact that large bodies of text were later crossed out leads to a heavily footnoted volume that is difficult to picture in the state in which Jefferson used it as a working project.

Several years ago a group of scholars decided to prepare a new and digitized version of this manuscript. Wonderful things have resulted from that collaboration, which is still in process. The scanning of the manuscript is being done in a manner that is superior to mere photographic capturing as would be done on microfilm. Moreover, one of the members of this research team came up with an idea that I doubt any "mere" historian or documentary editor would have imagined on his own. Taking a high resolution scan of one part of the manuscript that contained crossed out text, the technicians have done amazing things. Jefferson was quite skilled in crossing out text he wanted to obliterate from his writings and he did it in a fashion that makes it sometimes almost impossible (or always impossible without using advanced technology) to discern. With *Notes on Virginia*, Jefferson wanted to replace text he had written twenty or thirty or even forty years earlier, and he crossed some of it out with determination. A bright computer technician told the group that it might be possible to use image editing software to select the ink used to cross over old text and assign it a distinct color, say red. Then telling the computer to delete all red in the image would effectively obliterate the strike out, revealing the original black ink Jefferson had written years before. This is an amazing and wonderful development for scholarship, and it results from the wise decision to collaborate on this project rather than taking the low road of simply digitizing the text and getting it out on the Internet for the world to see.

Other collaborations may be somewhat more traditional but in the end enhance the results of the issue that concerns us. The Newberry Library decided fourteen years ago to prepare an Encyclopedia of Chicago. It was published this past October. From the beginning we wanted to prepare a manuscript that could be made available in digital form, and we had hoped to persuade our publisher (The University of Chicago

Press) to permit simultaneous publication. As our scholars got closer to completing the manuscript for publication, the Press kept shrinking the lag time they required between the print publication and electronic availability. It now is down to six months. But in the middle of this process our team, which was rather large, recognized that they simply did not have the resources to prepare an electronic database as they were finishing a manuscript for print. In the process of reaching that decision, we made what I consider to be great strides forward in our management of digital information. Our editorial team, led by three scholars, had in one of them a computer expert. We made an arrangement with the Chicago Historical Society to take our file and prepare it for availability on their server after the six month period. For the past several years this one specialized editor has been working with the Society to ensure the quality of that file and to make certain that the highest standards of digital information are maintained over time. The Historical Society, and not the Newberry Library, has taken on the responsibility for ensuring that this digital database will continue to be updated with new information as it remains available to world-wide users into the future. Collaboration, in this instance, therefore led to this focus by different teams, which may be viewed as separate work or as broader collaboration, depending on one's point of view. The Historical Society did not have to invest in creating the database, or in guaranteeing its reliability or thoroughness. Our team assumed that responsibility, and having the end product vetted by a university press before it went into print and then as the content of the soon-to-be available database, we feel that the result is the best possible.

Digital technology has made it possible to create huge databases that often force its creators to work in close collaboration. In these cases it seems to me that integration of talent and brain is made possible, or at least greatly facilitated, by technology. I note in my reading that I am seeing quotations from scholarly work, not always in digital form, attributed to two scholars who have worked together to create a piece of work. This is a relatively new but growing phenomenon, encouraged I believe by the new technology. We should note also that there is an incentive if not an imperative to grow large or massive projects. American Memory at the Library of Congress site comes to mind, and we are likely to see many more of these. Newberry staff were thinking of a regional project that would involve the digitization of images recently, and one of our consultants reacted to the suggestion that the project might encompass 150,000 images by saying "Well, you know, that's not so big." The argument was the larger the size of the file the lower the per unit cost. Such projects, fueled by technology, can become run away trains.

We at the Newberry welcome this new technology, and we like to think we embrace what we hope to be the highest standards in all that we do. Our staff is made up of scholars and professional librarians who view their work as closely integrated. We do not plan a scholarly institute without consulting the librarians who will be serving up our collections and helping the participants use them as part of the institute. The professional librarians, especially the curators, depend on the scholars to help them make informed decisions about acquisitions and about the effective use of our collections by scholars whose interests evolve over time.

The amount of data and information available to all of us today is almost incomprehensible. We as librarians must recognize the reality that young users of our materials think in greater numbers that everything is there somewhere on the Internet in digital form. The challenge to us is to keep these growing minds in a state of purgatory where a bit of intelligent skepticism is a good thing. No matter how much metadata of whatever quality is added to databases that we and others create, we must remain committed to the fact that scholarship is not a keyword search. It remains interaction with people through seminars, through research and consultation with people who can talk to a scholar in a way that leads to the discovery of things that neither might predict when the conversation begins. Much of the important work that scholars and libraries do is now in machine-readable form, like our catalogues and collection descriptions, but it has never been fully clarified even in the "old technology" of books and catalogues. Much important knowledge is imbedded in the *culture* of libraries and academic institutions, and much essential local knowledge is embedded in undocumented practices and the rich experience of sometimes quirky and irreplaceable individuals on our staffs. Standards and guidelines for our databases are necessary, but they are not sufficient by themselves. Librarians know LCSH and AACR2 and lots of other standards, but they also know how to apply them, or go beyond them judiciously, when confronted with novel situations. Good librarians also know *what* collections might relate to the half-formulated question a scholar is asking, even when the cataloging/metadata standards have never organized the collections in such a way as to make their relevance obvious for this purpose.

We should continue digitizing materials, with both eyes open, and both feet firmly placed in Purgatory. But we ought not to be simply digitizing page images to make texts readable, but (funding permitting) we should also aim to make the informational content of digitized texts machine-readable, so that we can ask and answer questions of works and collections in new ways, and for new purposes. Ultimately, scholarly and humane purposes must drive the whole enterprise; and furthermore, we are a long way from having any substitute for the human judgment required at nearly every stage of the process, from the formation of scholarly questions to the identification and use of appropriate evidence to answer those questions.

If we can achieve this in our brave new world of digital information and knowledge management, we will have created a high and healthy form of purgatory that some may think is quite next to paradise.

doi:10.1300/J111v46n01_06

Digital Libraries and Librarians
of the 21st Century

Nancy Davenport

SUMMARY. New forms of scholarship and publishing are radically and rapidly changing the relationships among those who create, store, distribute, and use information. This paper will focus on the changes in library collections and library information technology organizations, the resulting advancements in scholarly research, and will discuss the attributes, attitudes, and skills needed by the librarians of tomorrow to develop the strong connections between the academic disciplines and research libraries that are essential for library development in the 21st century. doi:10.1300/J111v46n01_07 *[Article copies available for a fee from The Haworth Document Delivery Service: 1-800-HAWORTH. E-mail address: <docdelivery@haworthpress.com> Website: <http://www.HaworthPress.com> © 2007 by The Haworth Press, Inc. All rights reserved.]*

KEYWORDS. Digital libraries, digital government, management of digital information, librarianship and digital information, staffing in the digital library

---

Nancy Davenport is President, Council on Library and Information Resources, 1755 Massachusetts Avenue, N.W. Suite 500, Washington, DC 20036 (E-mail: ndavenport@dir.org).

[Haworth co-indexing entry note]: "Digital Libraries and Librarians of the 21st Century." Davenport, Nancy. Co-published simultaneously in *Journal of Library Administration* (The Haworth Information Press, an imprint of The Haworth Press, Inc.) Vol. 46, No. 1, 2007, pp. 89-97; and: *Digital Information and Knowledge Management: New Opportunities for Research Libraries* (ed: Sul H. Lee) The Haworth Information Press, an imprint of The Haworth Press, Inc., 2007, pp. 89-97. Single or multiple copies of this article are available for a fee from The Haworth Document Delivery Service [1-800-HAWORTH, 9:00 a.m. - 5:00 p.m. (EST). E-mail address: docdelivery@haworthpress.com].

Thank you for the opportunity to participate in this conference. I wholeheartedly believe that these are some of the most exciting and challenging times for libraries and librarians and almost wish I were able to start a career all over again.

Some of you in the audience may know that I spent much of my career working with political bodies: the U.S. Congress, the National Legislatures of most of the countries in Central and Eastern Europe and with the United Nations. Thus it should be no surprise that I might look to the literature of political science for a framework for looking at digital libraries and the staff who design and operate them. I also decided that I should do the research for this paper online. I have two reasons to support that decision: (1) it's about digital libraries, let's see if it works; and (2) in some ways I operate in a fashion similar to any independent scholar or an information-seeking citizen. Unlike most of my colleagues who have spoken at this conference, I do not work in a library. I don't have a vast array of subscription databases at my fingertips. For someone whose most recent previous responsibilities were as the director of acquisitions for the world's largest library, I feel a bit like the cobbler's child.

In 2001, Sidney Verba of Harvard University was invited to deliver the Eckstein Lecture at the Center for the Study of Democracy on the Irvine campus of the University of California. His talk was titled: "Culture, Calculation, and Being a Pretty Good Citizen: Alternative Interpretations of Civil Engagement."[1] In his talk he posits that citizen participation in politics is a crucial component of democracy, and studying how citizens do participate has become a mainstay of political science analysis of democratic politics. He argues that the theories surrounding individuals as rational actors and cultural explanations of participatory behavior are insufficient to wholly explain citizen behaviors. I'm neither going to reargue or refute his position here today, but I want to apply some of his analytic framework to our focus, digital libraries. As Verba and his colleagues studied citizen participation they identified three sets of factors that foster participation: (1) being motivated to participate; (2) resources to invest in participation; and (3) mobilization. Now I'll try not to harm his research by distilling his scholarly and entertaining paper into a paragraph. Citizens participate in democratic life when they are generally interested in politics or when they believe their actions will have an effect and they want to influence government. In determining the resources required for the activity, they believed they had sufficient knowledge and time to participate and they were motivated to do so because someone asked them.

My second set of framing issues is drawn from the report by Jane Fountain to the National Science Foundation, "Information, Institutions and Governance: Advancing a Basic Social Science Research Program for Digital Government."[2] This report is the outgrowth of a 2-day workshop of more than 30 experts gathered at Harvard's Kennedy School of Government in 2002 to participate in the development of an agenda for digital government that was broadly based and multidisciplinary in nature. They started from the premise that though there were significant innovations in information and communications technology, digital government was at a very early stage of implementation and further that the implications of IT for the future of government were only dimly perceived in spite a "stream" of commentary, some informed, some speculative, and some suspicious.

To turn to libraries, the digital era has been evolving in libraries over the last 20 years. Librarians took the lead in much of the academic community when they converted card catalogs into searchable databases. Then adding electronic journals to their paper subscriptions, and replacing bibliographic instruction with information literacy training and now converting unique holdings to digital formats so others can have access for scholarly purposes. We have been present and persistent in analyzing the myriad issues in developing systems to support access and preservation. We have recognized the benefits that digital technology brings to academic inquiry and policy analysis, and we have incorporated them at every possible point.

How would this be explained in Verba's terms? Where are the motivation, the resources, and the mobilization? In looking at our profession the motivation comes from multiple hallmarks: to be active partners with the scholars on our campuses, to ensure that students have the best resources at their disposal and the skills to engage them for lifelong learning, and our innate predisposition to simply push information at people. The resources to participate in digital library efforts could be looked at in two classic definitions of supply and demand. For supply, we have financial resources to purchase or lease new information sources, but more important to my mind are the collections of unique cultural materials that already exist and we have the staff that want to work in this new exciting field and curate converted collections. For demand, we have to look no farther than the typical college freshman who has grown up with a mouse or a joy stick always at hand and who merge the typical definitions of work and play as they swirl through cyberspace doing both.

But I think the most interesting of the factors to explore is mobilization. Another perhaps innate characteristic of librarians is collaboration or cooperation. We belong to the same associations and societies and many librarians have long lasting personal friendships that grew out of professional meetings. When we decide to accomplish a task as enormous as creating digital libraries, there are colleagues to consult, there are associations or organizational umbrellas to shelter the project and if there are not the right ones, we'll create them. And admittedly, another mobilizing factor has been the ability to secure funding for particular projects from federal agencies or foundations.

Now let me return to the social science research program for digital government and see how it relates to our digital libraries. The workshop participants at the Kennedy School presented a research agenda that focused on the intersection of IT, organization, and governance. They developed four strategic areas for research: (1) Information Technologies, Governance and Organization; (2) Digital Government and Its Stakeholders; (3) Change, Transformation, and Co-evolution; and (4) Systematic Research Design. Allow me to give you a couple of examples of the research questions they posed for digital government. How does IT interact with the structure and processes of government organizations? How are government managers and policy makers using IT to develop new organizational forms or to modify existing forms?

Now let me translate those and other provocative ideas they posed:

- How does IT interact with the structure and processes of higher education? Of my university? Of my library?
- How are university managers and policy makers using IT to develop new organizational forms or to modify forms?
- What are the impacts of IT on inter-institutional coordination and collaboration?
- What policy and processes influence data integration and standards? How do they do it?
- How do students actually use online information and services?
- How are interest groups, scholarly societies, and student groups using the Web?
- What are the key emergent changes that might be empirically identified and described in scholarly engagement?
- What is the impact of increasing use of information-based, networked forms of organization on the institutional structures–for example, oversight, budgeting, and accountability systems–that regulate university life?

Applied to digital libraries in the academic world, these questions could be a compelling research agenda.

I think we have an opportunity at hand to begin to look at a part of this agenda, to rethink professional education and continuing education, where I believe we have placed obstacles in the path of change and how we define a librarian. And we need to create more points of entry into the profession.

The Institute for Museum and Library Services has commissioned a study that will help inform the curriculum of what the librarian in the 21st century ought to have as competencies upon completion of the Master's degree. The principal investigator for the student is Dr. Jose-Marie Griffith, dean of the Graduate School of Information Studies at the University of North Carolina. Dr. Griffiths has formed a national advisory panel to guide the study and CLIR is respresented by Susan Perry. We were especially happy to see this study be announced.

It is my proposition that graduate education has not kept pace with the needs of the scholarly community. Or to say it in a more collegial way, the educational needs of librarians in order to be expert practitioners in their field have changed multiple times and the rate of change is increasing. And the schools that educate librarians have not kept pace. Or to restate that in a more collegial way, we are expecting the graduate school program to prepare both entry level librarians and senior managers at the same time and through the same course work.

My second point is that we need to create other entry points into the profession. Much of the work we do, particularly in academic libraries and in digital curation, requires much more in-depth discipline specific knowledge than we have needed in the past. I would make a third point: I would challenge us to realize that the entire profession has to be open and welcoming of colleagues with other competencies or we have to re-dress what constitutes library and information science.

We'll explore this idea by examining a series of library settings that require very different skills than are being taught in many graduate school programs. And begin by looking at the paradox of libraries today.

A small liberal arts school in Ohio doesn't have any staff in a librarian position. They hire librarians but the positions are described as library-technology consultants. An LTC, as they are called, could be working with a professor to set up the electronic reserves for next semester's classes, or teaching a class in Dreamweaver or reconfiguring the college's firewall. This college has combined the library, academic computing and administrative computing and assigned all the managerial responsibility to the college librarian, now the Chief Information Officer. All technology used in the college's teaching and learning programs, and even the telephone system are the library's responsibility. One could ask why the assignment went to the library instead of to the former heads of academic computing or administrative services. The answer is that the library understood information management for the college-wide perspective and combined that intellectual framework with the service attitude that is a hallmark of the library professional. The librarians were able to think holistically regarding the automation needs of the campus.

To give another example, after attending a conference sponsored by the Council on Library and Information Resources (CLIR) on managing digital assets, one of the participants returned to his college and created a new position. The new position, the Digital Initiatives Librarian, will have responsibility for managing the college's digital assets. So, rather than filling an existing open position and hoping for the right skills to be attracted to the job, he acquired new insight on what expertise was needed in his institution and the vocabulary needed to describe the position so that the position can attract the candidates with the right skills for a job they can be excited about.

Another example: Several scholars at the University of Virginia have created a remarkable piece of digital scholarship entitled the Valley of the Shadow, a study of two towns in the Shenandoah Valley during the Civil War. One town was located at the northern end of the Valley and did not hold slaves. The other town was at the South end, in a slave holding state. Their scholarship brings together every scrap of evidence that documents the social, demographic, and economic conditions of the two towns before, during, and after the Civil War. The working relationship has been a trusted one among the scholars, their department, and the university library. But now one of the scholars is moving to another university. He has announced that he intends to create the same kind of digitally based scholarly work that will focus on the settling of Omaha, NE. But the new site will have to draw on some of the underlying programming on the Virginia site. The Omaha site will parallel the Shenandoah site. This scholar has just crystallized the dilemma: he is totally reliant upon the two libraries to keep his work alive, in parallel form. What kind of guarantee might he ask for? What kind of guarantee could the librarians offer to him?

Another example: One of the most powerful uses of digital technology is to share scarce resources. Libraries have been digitizing their special collections. They have done this individually, in joint operations, and as a group created to do exactly that. But as the digital surrogates are created and exhibited, a new need arises: the need for curatorial skills and subject knowledge. The collection created has to hang together. It has to tell a story. It has to exhibit scholarship in sufficient depth that collection development and curation, collection description and metadata, and collection use have to work seamlessly. It has to reveal itself to the user who stumbles into the Web site more directly than to the scholar who seeks the item or the entire site.

What do these few examples tell us about the kind of knowledge and the kind of skills that librarians have to have in today's academic library?

At the liberal arts college in Ohio, we see the need for a director with vision to guide the integration of major programs within the college, ones which in some settings have been strong competitors for resources. In addition to vision, the director has to see how the pieces can make a programmatic whole and have the management skills to fashion the new organization that is stronger than the contributing slices. The librarians working in the merged organization need several equally strong sets of competencies: a fluency in technology and its applications to the world of teaching and learning, teaching skills to create and deliver a course to a classroom full of students, research skills to work with and support faculty in their research and teaching initiatives.

The librarian creating online collections has a different set of needs: many of the attributes of a museum curator are called for including deep knowledge of the underlying subject matter, contextual history, and preservation talents.

The libraries charged with keeping digital scholarship of its faculty members available for research not only have to have the technology and the staff to do so, they have to have the will and the knack of collaboration.

None of the examples I've given are out of the ordinary. Each institution I've described is a real place and all people are talented colleagues.

We see plainly that new forms of scholarship and publishing are radically and rapidly changing the relationships among those who create, store, distribute, and use information, and that change is happening first in the academic community.

Libraries that have had large serial collections now have licenses to use or to use and archive large numbers of electronic serials. Multiple issues are solved by delivering e-content: all the cumbersome steps of managing paper collections from check-in to tracing missing issues, reshelving materials after use, protecting them from mutilation and theft and preservation binding. The very supports of the building should be sighing with relief that the enormous weight of serials collections is not being added to.

In a 2004 study published by CLIR that researched the nonsubscription side of periodicals, changes in library operations and attendant costs were compared between print and electronic formats. They also note that on campuses across the country digital technology is changing the behaviors of both students and faculty as they seek information, even in disciplines rich in the world of print such as history and literature. In the 10 or so years that libraries have been making the shift to electronic resources, there have been conservative and exaggerated predictions about the cost impact of such a change in format and in services. Some promised it would save oodles of money; others foresaw the possibility that business would be done as usual and that while cost might be higher, the benefits to the user combined with the streamlined processing made the change worth it at almost any cost. Librarians also saw new ways of cooperating in purchasing or at least in negotiating and the consortiums that came to life during the same period are the equivalent of professional buying clubs. But often when faced with a big deal purchasing option, we opted to buy in e-format titles we would not have included previously and would have refused.

What have been some of the staffing changes? For one, few manual labor tasks for which library technicians would be the appropriate level. Professional librarians still most often do even the cataloging of e-serials. The additional tasks that were added were the management of licenses. Technology specialists were added to assure that the machines were configured to comport with the terms of the license, and to run tests that reassured us that data was not corrupted. Lawyers were added to the mix in ways never seen in the old model of pay a subscription and get 12 issues. Suddenly we were all worried about who or what was authenticated or indemnified.

Scholars and librarians are forming new intellectual partnerships. Creators and publishers of scholarly resources are seeing how their business decisions–not just those of libraries and archives–can influence or impede access to information resources in the future. Libraries offer vastly different content, research methodologies and services for their faculty members and students than they did 15 or even 10 years ago. They are organized and operate differently and frequently more collaboratively than in the past.

Today's information professionals require new skills and expertise to work effectively in an environment characterized by rapidly evolving technologies and organizational structures to cope with the diverse demands of information seekers.

The Council on Library and Information Resources is actively studying, publishing, and testing assumptions about the Librarian of the 21st century. We believe our role is to be a catalyst–to move the scholarly communications process forward, to identify needed linkages between academic disciplines and research libraries and the traditional and digital information resources that support them, to help prepare information specialist with the tools they'll need. We are particularly interested in working with faculty and information technologists who collaborate in building digital resources for scholarly research in teaching. In addition to the research, we are prototyping a program that proposes a different route to academic librarianship. This model requires deep knowledge of a subject area in the humanities, an intensive orientation program in a seminar setting, and close supportive mentoring by a librarian well established in an academic setting.

In conclusion, I've been very excited by the developments in academic libraries as I traveled to campuses over the last six months. When we think broadly about the libraries we are developing today for the new crop of students and scholars, and about the developments announced by Google for the digitization of 19th century materials, it appears that we will distinguish our institutions one from the other by the services we can deliver applying academic inquiry tools to digital collections. It's not too soon to decide that we don't all have to study cataloging or memorize 200 reference sources.

Thank you for the invitation to participate.

NOTES

1. Sidney Verba, "Culture, Calculation, and Being a Pretty Good Citizen: Alternative Interpretations of Civil Engagement." (2001). Center for the Study of Democracy. Paper 01-01. http://repositories.cdlib.org/csd/01-01.

2. Jane E. Fountain. Information, Institutions and Governance: Advancing a Basic Social Science Research Program for Digital Government. Harvard University. RWP03-004 https://ksgnotes1.harvard.edu/research/wpaper.nsf/rwp/RWP03-004/$File/rwp03_004_fountain.pdf.

doi:10.1300/J111v46n01_07

Toward a Topography
of Library Collections

Gary M. Shirk

SUMMARY. This paper explores some fundamentals of human perception, pattern recognition, concept formulation, and knowledge storage which suggest new opportunities to understand library collections and develop collecting strategies. Building from these principals, the author demonstrates how advances in digital technology may make it possible to develop a topography of library collections. doi:10.1300/J111v46n01_08 *[Article copies available for a fee from The Haworth Document Delivery Service: 1-800-HAWORTH. E-mail address: <docdelivery@haworthpress.com> Website: <http://www.HaworthPress.com> © 2007 by The Haworth Press, Inc. All rights reserved.]*

KEYWORDS. Perception and information visualization, visual imagery, visualization of library collections, pattern recognition, concept formulation

INTRODUCTION

Describing and analyzing library collections and collection development has a long and distinguished history, but has, for the most part, re-

Gary M. Shirk is President and Chief Operation Officer, Yankee Book Peddler Library Services, 999 Maple Street, Contoocook, NH 03229 (E-mail: gshirk@ybp.com).

[Haworth co-indexing entry note]: "Toward a Topography of Library Collections." Shirk, Gary M. Co-published simultaneously in *Journal of Library Administration* (The Haworth Information Press, an imprint of The Haworth Press, Inc.) Vol. 46, No. 1, 2007, pp. 99-111; and: *Digital Information and Knowledge Management: New Opportunities for Research Libraries* (ed: Sul H. Lee) The Haworth Information Press, an imprint of The Haworth Press, Inc., 2007, pp. 99-111. Single or multiple copies of this article are available for a fee from The Haworth Document Delivery Service [1-800-HAWORTH, 9:00 a.m. - 5:00 p.m. (EST). E-mail address: docdelivery@haworthpress.com].

lied upon expert opinion to identify gaps or deficiencies. Attempts to use statistical analysis to gain insights have been productive, but they require readers with considerable prowess in abstract thinking and a degree of mathematical patience that eludes most of us. I know that Bob Nardini, YBP's Chief Bibliographer, and his colleagues Tom Cheever and Charles Getchell, will pardon my reference to their 1996 article, "Approval Plan Overlap: A Study of Four Libraries."[1] Their detailed and extensive tables contain useful information, trapped in a presentation that requires more time to assess than most of us would be willing to give it. It could have been worse. They were careful to avoid statistical tools that would further alienate all but the most dedicated reader, and their insightful narrative helped us to overcome the obstacle of all that information. However, we are left wishing that we could peer through the numbers to see where overlap existed and where it did not. This study, like so many others, relies upon our conceptual abilities and underplays our senses.

The near ubiquity of digitized information offers much more than electronic text, eJournals, online reference, networked campuses, and the WWW. It also offers the opportunity to gain insights into our existing library collections, both digitized and print, and into the development of our collections over time, as single institutions or in groups of institutions. Technologies that employ digitized information about the content of these collections make it easier than ever before to understand extremely large data sets.

These technologies, explored in the emerging field of Information Visualization, harness our innate abilities to grasp complicated images, sometimes involving thousands of data points. This paper doesn't attempt to summarize the field for library professionals. Fortunately, the January/February 2005 issue of *Library Technology Reports* has already done so.[2] Rather, my paper explores why information visualization works for us and why it could help us to see features of our collections and collecting patterns as clearly as we see geographic features of a topographic map.

PERCEPTION AND INFORMATION VISUALIZATION

In his fascinating little book on the development of Microsoft's ClearType, *The Magic of Reading*, Bill Hall begins by quoting Stanley Morrison, a noted typographic historian, "What they call Originality is achieved by getting down to the root-principle underlying the practice.

From that origin you think your way back to the surface, where you may find you're breaking untrodden ground."[3] Appreciating fully why information visualization helps us to understand large, complex data sets, requires that we spend some time with root principles. By challenging our assumptions and exploring fundamental concepts, we can learn to see as new what has become commonplace.

Nothing could be more common than how we perceive the world. It just happens, like rainfall on a summer day. But hidden in this everyday experience is our only means of reaching out to the world, of comprehending what we see, touch, and hear. Simply put, our understanding of any complex phenomenon is limited by our ability to perceive it and then to our success in learning from that perception. Our five senses (seeing, hearing, feeling, smelling, tasting) are the ways in which we learn about the world. Relative to other animals and insects, these senses are very limited. For example, compared to the falcon, we are myopic; compared to the bee, we see only a limited part of the light spectrum; and compared to the family dog, whose sense of smell is 100,000 times as sensitive as our own, we walk around with a perpetual stuffy nose. So it goes with our other senses. None ranks supreme among animals and yet they are the doorway to the world around us.

Of our five senses, we have come to depend most upon sight because vision allowed our ancestors to see the approach of predators at a distance and, at the same time, allowed us to perceive the tracks of our prey. Depending upon our sense of hearing to detect the stealthy tiger would have been risky and being in close enough proximity to smell, taste, or feel it, would not likely have been an enjoyable learning experience. With this in mind, Bill Hall, writing in *The Magic of Reading*, concludes that we humans are sight dominated, serial pattern recognition masters who owe this skill to our prehistoric ancestors.[4] Presumably, our ancestors whose eyesight and visual cortex were best adapted to clear sighted observation and tracking survived; others, who lacked these abilities, did not.

Visual perception is not simply a mechanical response to light. It is a combination of the response to light and the brain's cognitive processes. Both must be present. While it can be argued that cognition and the ability to form concepts exist even in the absence of what we sense, it's indisputable that what we learn must have its origins in what we can perceive. As a result, we seem to understand best that which is most accessible to our unaided senses. We see patterns easily when they are fixed, but only weakly when they are in motion because visual and cognitive systems are tuned to speeds most useful to our ancestors. If motion is too fast or two slow, we cannot perceive it. Can we see a bullet in flight or the beat of a hummingbird's wings? Can we see a flower bloom or a glacier move? While we sense less about the world than others that share our planet, we have learned more about it than any other animal. How is this possible?

We have learned to build upon what we sense by creating devices, external to ourselves, which extend our perception and enhance our cognitive abilities. Much of what we accept as the common trappings of modern life are such artifacts. Devices such as the telescope and microscope extend our sight as does the more common pair of eye glasses. All of these extend our appreciation of the visual spectrum. We use other devises, like the infrared camera, to transform spectra that we cannot see to visual light that we can. To appreciate the role of how we use the external world to enhance cognition, try the following experiment suggested by Card, MacKinlay, and Schneiderman in *Readings on Information Visualization.*[5]

First, multiply to find the product of 53 times 75 in your head without recourse to pencil or paper and record how much time it takes. Now try repeat the problem using pencil and paper:

```
    75
  ×53
   225
   375
  3975
```

As Card and colleagues observe, "As this informal demonstration shows, visual and manipulative use of the external world amplifies cognitive performance, even for this supposedly mental task."[6] Of course, had the problem been a little more complicated, the time to reach a solution would be even greater. If it had been still more complex, few of us would have been able to solve it in our head at all.

What is the process behind this? Basically, memory studies suggest that people can maintain only a few items in memory over the short term. A complicated sequence of math stretches this memory limit. Using pencil and paper extends the limit because we're able to store part of the information outside our memory. In addition, we manipulate the visual space to structure the problem to make it easier to solve, e.g., by setting the second line of the multiplication one character to the left to represent 3750 prior to addition. By recording the problem visually, we reduce the information that must be active in short term memory, speeding up the process.

Likewise, we find that the use of rehearsal and visual imagery helps to extend our long term memory which is one of the reasons that memory experts emphasize the use of images to assist learning facts and concepts. Perhaps this is why our prehistoric ancestors first daubed painting on cave walls to aid their memory of important events or why, in the dawn of civilization, great stone monuments were built to calculate and commemorate the seasons. As our concepts become more complicated, we rely more on external cues to aid memory and enhance how we learn and store knowledge at the borders of our ability to perceive and understand them directly.

Graphical representation of complex, large scale data is not new nor is it peripheral to our ability to comprehend the world around us. As useful as it is to understand phenomena that we can perceive, it is far more useful to help us to understand what lies outside our perception, whether real or abstract. All of us have seen a model of our solar system, but none of us have seen the solar system itself; we've all learned to understand DNA from Watson and Crick's model of its double helix structure, but none of us have seen DNA itself. Such models whether depicted in two or three dimensional space are essential to understanding these phenomena. So it is not difficult to see how valuable visualizing information can be to achieving a similar understanding of abstract phenomena.

Edward Tufte's seminal work, *The Visual Display of Quantitative Information*, developed a theory of graphic illustration which underpins much of the work in information visualization today. Tufte argues that excellence in graphic presentation requires that complex information be conveyed with clarity, precision, and efficiency. The best graphics convey "the greatest number of ideas in the shortest time, and the least ink in the smallest space."[7] In sum, excellent graphics enhance cognition with the least amount of effort.

In a book filled with fascinating examples–good and bad–of attempts to use graphics to enhance understanding of complex issues, one of Tufte's examples struck me in particular. This was the Yu Chi Thu (Map of the Tracks of Yu the Great) of China dating from approximately 1100 AD and displaying a precision unequaled in Europe for about 500 years.[8] The map's designer used a small grid drawn on the illustration to illustrate a much larger space in the real territory. What a leap of intellect this must have taken the first time this technique was used. I'm sure the author of the map was not the genius who made the discovery. That must have come much earlier otherwise the map's creator would not have had a community of people who could make any sense of it at all. This early map illustrates an important principle of graphic illustration. To be effective, it requires a set of conventions adopted by a community of people with a common ability to perceive visual display.

As big as this leap of intellect might have been, using a distance scaled on a sheet of paper to represent the passage of time is much more so. In the real world, the world we experience every day, time and space are two very different dimensions. No one would equate an inch to a second or a mile to an hour. One is clearly a matter of physical space while the other is one of duration. Yet, we easily find ourselves saying–or at least up in New Hampshire we do–saying that Boston is about an hour and a half away, meaning that it requires a specified passage of *time* to travel that *distance* by car. Nevertheless, we have to give respect to that genius long ago who realized that a unit of time could be illustrated as a unit of distance measured on a sheet of paper. Tufte provides some excellent examples of how this illustration has been used and abused over the last few centuries.

Of course, this ingenious device then begs the question, if a visual representation of distance can be used to illustrate an abstract like time, could other visual effects be used to depict other abstract dimensions? Might these effects be used to show dimensions which we can neither experience directly nor perceive at all? According to Tufte, it took fully 5,000 years after the first geographic maps were drawn on clay tablets for advances in cartographic and statistical skills to come together enough to depict the first data map, i.e., a map that includes non-spacial information in graphical depiction.[9]

In Edwin Abbott's classic, *FlatLand: A romance of many dimensions*, we learn about the residents of a two dimensional space world as they discover a third dimension.[10] The residents cannot perceive that dimension but they can infer its presence from what they can perceive. It was written to give Victorian era people the ability to understand time as a 4th dimension, a dimension on which change occurs as much as it does in three dimensions–even if we cannot perceive it. Today it seems quaint, and the inter-dimensional experience that Abbott depicts seems tame to generations brought up on Star Wars films. But, in a companion text that pays homage to Abbott's work called *Flatterland: Flatland only more so*, Ian Stewart finds a way to shock this generation as Abbott did his. He shakes us out of a tidy 3 or 4 dimensional world in which we've become reasonably comfortable and spins a story to help us understand a multidimensional world.[11]

Stewart argues that while everything we perceive and directly experience trains us to think of the world in terms of three or maybe four dimensions, the world we inhabit is N dimensional. But most of these "extra" dimensions, like those at the sub-atomic level, cannot be perceived. Instead, we rely upon the quantum physicist's mathematical proofs and arcane experiments to deduce their existence. Most of us would not call these dimensions. For us, the word dimension connotes a direction in space, so we're more likely to call these variables, characteristics, elements, factors, or aspects. Nevertheless, these dimensions are as much a part of our reality as other phenomena we cannot directly perceive.

In the same way that we transform part of the invisible light spectra to a range that we can perceive, we can use graphic illustration to transform invisible or abstract information. For example, it's become a convention to illustrate hot objects using the color red and cold objects the color blue or green though no one would confuse the color spectrum for temperature variation. Using such techniques, we have learned to portray multiple dimensional phenomena in two or three dimensional space.

One graphic display which we all recognize is the topographical map. But its commonality masks what a marvelous device it is for displaying thousands of data points in a visually digestible way. Using such maps, we can illustrate the entire world at a glance: contours, shapes, elevations, and relationships of one surface feature to another. These are all spatial dimensions. Imagine the thousands of latitude and longitudinal coordinates required to outline continents and the thousands of elevations needed to show height above sea level. Yet all are easily discernable by transforming measures of elevation to variations in shading or color. To this map we can add other conventions, such as political boundaries, earthquake faults, or population–all depicted with line, or color, or shapes applied to the map.

How is it that maps like these can convey so much information, thousands and thousands of bits of data, such that we can immediately, almost intuitively comprehend it? These maps rely not only on our innate ability to comprehend complex images, but also upon a standard set of conventions: north/south, east/west, scale, color for height or shade of gray, grid lines, etc. The conventions, like the spatial organization of the math problem on paper, assist by providing a framework in with new information can be organized efficiently. It is the language of maps, the fundamental structure that permits instant recognition and makes communication possible. Combined with a strong visual graphic, the maps trigger instant recognition and convey unmatched breadth of information in an instant.

TOWARD A TOPOGRAPHY OF LIBRARY COLLECTIONS OR COLLECTING

Today, information visualization techniques and software provide the opportunity to visualize and manipulate multi-dimensional data sets collections in ways that have never before been available. We can now provide a multi-dimensional display of whole collections or a year's worth of acquisitions, or the acquisitions of one library compared to another, or even the acquisitions of an entire consortia of libraries. But, does the ability to create such visualization mean that we should take the time to do so? What would be the benefits worth of such an effort?

Like a topological display of the world, information about our collections or collecting practices visualized in its entirety may provide useful insights about how the collection compares to other collections, how it is developing over time, unintended gaps in the collection, strengths and weaknesses in the support of interdisciplinary studies, the extent to which collection contributes to the whole of the consortia in which you participate, the scatter of material needed for one discipline across the full subject scope of the library, and so on. The value is that relationships hidden from view just 35 feet up become crystal clear at 35,000 feet above. Current collection assessment largely function at 35 feet up, at the level of title lists and subject-by-subject title counts. While these are often supplemented with statistical measures of collection totals, they do very little to show the collection's overall shape and texture.

It takes time to make sense of statistics, about collections, or anything else. Sometimes, they can obscure reality as much as they illuminate it. For example, answer this question: how is it that a young, athletic swimmer from Arizona drowned in a river whose average annual depth was just six inches? You might be at a loss if only given the statistic of average annual depth. However, given a graph of river depth over the 52 weeks in a year, you'd know the answer immediately and would probably be able to pin point the weeks in which the young man might have died. In the same way, statistics showing the percentage of total books collected in a subject doesn't necessarily tell you if the correct books were acquired or provide any information on the interrelationships of many of its dimensions.

Information visualization tools and software permit multi-dimensional display depicted in two dimensional space, such as the topographical maps we saw earlier, or even a virtual depiction of three dimensional space. Experimentation in 3D visualizations run the gamut from Cityscapes to concept mapping, from landscapes to perspective walls, from topic maps to trees.[12] The body of research devoted to explorations of the use of information visualization to reveal library collections is not large, but it is growing. One of the more promising approaches is the use of "starfield" representation, which plots titles on a two dimensional grid, then uses various filters to zoom in and out of user selected areas on the grid.[13] In another approach, Schneidermann and his colleagues apply a hierarchical browser with a two-dimensional visualization that enables users to explore large search results rapidly and easily.[14] Both of these efforts, like most in information visualization to date, have been the computer science research and have not been driven by librarians.

However, information visualization software may have reached the point that we practitioners can now provide some leadership without having to learn programming. We may want to begin experimenting by reviewing Websites such as Olive: On-Line Library of Information Visuzalization Environments, www.otal.umd.edu/Olive/; Highwire Press TopicMap, http://higwire.standord.edu and others recommended in the January/February 2005 Library Technology Reports; then learning more by testing currently available software products such as Grokker, www.groxis.com/sservice/grok or Rooms, http://rooms3d.com.[15] For something more immediately accessible we might spend some time with Spectra CRC by Library Dynamics, www.librarydynamics.com. SpectraCRC is an interesting tool for analyzing collections that employs some information visualization technologies. It's a promising start that illustrates the value of graphic displays of multiple variables such as title counts by subject and circulation.

Despite the progress that's been made in information visualization, issues of presentation and functionality remain important. For example, Library Technical Reports lists nine functionalities that should remain in the forefront of information visualization design, of which the following would be most important for a topography of library collections and collecting:

- Close connection to the data through zooming and probing features;
- Multidimensional approaches to displaying and manipulating multiple attributes;
- A picture-centric user interface that is easy to use and understand;
- Control of the underlying data model through manipulation of visual results, which supports iterative, efficient investigation of data;
- Interactive control of level of detail and resolution; and
- Ability to combine visual queries with visual data mining to form "What if" questions.[16]

To date, the designers of this information visualization software have been most concerned with the Graphical User Interface (GUI), i.e., graphical design and user experience of on-screen tools. Everyone is seeking the ideal manifestation of "the visual information seeking mantra: overview first, zoom and filter, then details on demand."[17]

As necessary as this functionality might be, it is not sufficient to build an intelligible topography of library collecting and collections. Dazzling visual displays can do little to add information to data sets devoid of meaning. To make the visualizations useful communications tools, more attention must be given to both data and the conventions that structure them.

IMPORTANCE OF CONVENTIONS

In order for such recognition and communication to take place, a community must share both a common ability to perceive the world outside themselves and a similar set of broadly held conventions. These two elements are not only essential in communication itself but also in how we learn, store knowledge, recall it, and put it to practical use in our daily lives. In short, we think as we perceive and communicate. So, we can only comprehend and put into practical work what we as a group can see and share with one another.

Conventions, unlike scientific truths or theories, are not discovered or reached through a process of logical deduction, rather they are principles reached by a process of agreement or achieve such agreement through longstanding use. Our conventions are everywhere around us, so commonplace that we no longer recognize them as mere conventions. Some conventions have been elevated to the status of laws and regulations by a formal process of legislative agreement. Others have evolved over time to acquire the aura of natural law. But, fundamentally all have reached this level by a process of agreement. Language, both written and oral, our customs, road signage, educational focus, legal practice, etc., are conventions, the framework that guides how we interpret the world from our community's perspective.

We in the library profession possess important conventions for the description of works in collections: Dewey, LC, Subjects Descriptors, cataloging rules, etc. But we have few conventions for the visual display of whole collections or even parts of collections. In essence, we have taxonomies, classification systems and descriptive guides, which are used to describe individual titles and provide the means to organize groups of titles in a reliable scheme. Advances in computing and the near ubiquity of digitized meta-data make it possible to gather, analyze, and display information in ways that would enable us to evaluate multivariate data for total collections at a glance, much the same way as we do topographical maps of the world. While the tools exist to do this, we lack the broadly held conventions that computer generated topographical maps borrow from hundreds of years of print development.

While we can look to graphic designers and computer software engineers to design display software needed for library topography, we must look to ourselves to establish the conventions that will govern the meaningfulness of the information displayed. We must develop conventions beyond the accepted meanings of cataloging descriptors to build consensus for other metadata that may be useful in multi-dimensional displays of collecting patterns.

Descriptive information such as that contributed by approval plan vendors, would be a good starting point. These have a long history with proven utility beyond traditional bibliographic description. Descriptors for readership level, interdisciplinary treatments, geographical origin or coverage, specialized aspects, chronological range, aggregate sales, price, etc., would all be useful dimensions in a data set. In total, the number of dimensions that may describe a work could be numbered by the dozens. However this information has been collected by individual firms and has been shaped by these firms for their own and their customers' benefit. Between the vendor and its customers, these values have the substance of well-accepted conventions, but, for the most part, this strength is lost to the profession as a whole.

While we might have generally accepted definitions for some of these, we certainly don't have much agreement on the values that might be applied to them. One of the more controversial ones is Readership Level for which YBP applies four values: Popular, General Academic, Advanced Academic, and Professional. Why only four? Why as many as four? What do we mean with each of them? How are they applied to titles? While we have an operating definition of these at my company and tested criteria for their application, we do not, as a profession, have conventional meanings for them. The audience for any topography that would use them must be limited then to the group for which they are acceptable conventions. Without consensus agreement to their meaning, any topography of library collecting would provide little true information and may lead to confusion.

Information visualization of library acquisitions and the subsequent collections is therefore hampered on two counts: (1) the number and strength of existing conventions is narrow, and (2) there's virtually no consensus or agreement on how these conventions might be depicted visually. Imagine how difficult it would be to create a topographic map if there were no conventions governing scale, shapes, contours, distance relationships, and height to govern the data. Further, it there were no traditions of representation to govern how that data is transformed into visual analogs and displayed as a map, how would anyone comprehend the display? The lack of meaningful conventions is therefore the single most important obstacle to library collection topography.

CONCLUSION

Information visualization technologies combine the ability to manage large data sets, graphic display, and interactive tools that allow analysts to view the whole, then drill into the details using easily manipulated on-screen tools. The effectiveness of these technologies derives from our innate abilities to grasp complex visual images instantly and recognize the patterns that might be displayed. The meaningfulness of the displays, however, depends upon the fundamental set of conventions that give meaning to the graphics. The search for the conventions that will ultimately give shape to collecting landscapes is too important to be left to graphics designers and information specialists. Only when we practitioners get involved in defining our conventions will effective graphics be developed and a true topography of library collections evolve.

NOTES

1. Robert Nardini, Charles Getchell, Jr., and Thomas Cheever, "Approval Plan Overlap: A study of Four Libraries," *The Acquisitions Librarian*, Number 16, 1996: 75-97.

2. *Library Technology Reports*, Jan/Feb 2005, Vol. 41, Issue 1.

3. Hill, Bill. *The Magic of Reading*, Microsoft Corporation, eBook: 13.

4. Hill, p. 19.

5. Stuart Card, Jock Mackinlay, and Ben Schneiderman. *Readings in Information Visualization: Using Vision to Think* (San Francisco: Morgan Kaufmann Publishers, 1999): 1.

6. Card et al.: 1-2

7. Edward R. Tufte, *The Visual Display of Quantitative Information* (Connecticut: Graphics Press, 1983), p. 51.

8. Tufte: 20.

9. Tufte: 20.

10. Edwin A. Abbott, *Flatland: A romance of many dimensions,* 1884.

11. Ian Stewart, *Flatterland: like Flatland, only more so* (Cambridge: Perseus Publishing, 2001).

12. *Library Technology Reports*, Jan/Feb 2005, Vol. 41, Issue 1: 14.

13. J. Alfredo Sanchez, Michael B. Twidale, David M. Nicols, Nabini N. Silva, "Analyzing library collections with starfield visualizations," Working Paper 09/2004, Hamilton New Zealand: University of Waikato, Department of Computer Science, 2004.

14. Ben Shneiderman, David Feldman, Anne Rose, and Xavier Ferre Grau, "Visualizing Digital Library Search Results with Categorical and Hierarchical Axes," University of Maryland, Department of Computer Science.

15. *Library Technology Reports*, Jan/Feb 2005, Vol. 41, Issue 1: 57-58.

16. *Library Technology Reports*, Jan/Feb 2005, Vol. 41, Issue 1: 16.

17. Shneiderman et al.

doi:10.1300/J111v46n01_08

Index